TI-83/84® PLUS MANUAL

SUSAN HERRING
Sonoma State University

INTRODUCTORY STATISTICS
NINTH EDITION

ELEMENTARY STATISTICS
EIGHTH EDITION

Neil A. Weiss
Arizona State University

Addison-Wesley
is an imprint of

ISBN-13: 978-0-321-69149-1
ISBN-10: 0-321-69149-0

1 2 3 4 5 6 BRR 15 14 13 12 11

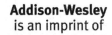

Addison-Wesley
is an imprint of

www.pearsonhighered.com

PREFACE

This manual is designed to aid the reader in using the TI-83/84 Plus graphing calculator to solve statistical problems. It is not meant to show the only way to solve these problems. Specifically, the manual is designed to be used with *Elementary Statistics*, 8th Edition and *Introductory Statistics*, 9th Edition by Neil Weiss (Boston, Massachusetts: Pearson Addison-Wesley). However, it can be used to supplement any textbook covering basic statistical topics.

It is expected that the student will learn and understand the reasoning behind the computations and that the TI-83/84 Plus will then be used to decrease the time spent on computation and increase the time spent on statistical reasoning and analysis. Most of the basic statistical computations are built into the TI-83/84 Plus. For some of those that are not, programs have been written using the other capabilities of the TI-83/84 Plus. These programs are optional and are contained on the WeissStats CD. Also, the TI-83/84 Plus has random number generators for several distributions that can be used to model random samples and further explore statistics. However, it must be remembered that the TI-83/84 Plus is not a computer and although it is very capable, it does have limits on the amount of data and programs stored in its memory.

The manual begins with a Preliminary Chapter that is intended to introduce the student to the basic commands of the TI-83/84 Plus. The chapters then follow the development of topics in the texts, making it very easy to use as a supplement. Each procedure is demonstrated with an example from the textbook illustrating how to use the relevant TI-83/84 Plus procedures. The examples and exercises in this manual are numbered to correspond with the examples and exercises in *Introductory Statistics* making it is possible to compare the textbook solution with the solution obtained using the TI-83/84 Plus. Additionally, some sections of the manual are not covered in *Elementary Statistics* and these are noted with an asterisk (*).

I want to thank Neil Weiss for writing two first-rate textbooks and for allowing the use of textbook examples and exercises in this manual. I would like to thank all the people at Pearson Addison-Wesley for their efforts and support. I especially wish to thank Dana Jones Bettez and Joe Vetere for their direction and advice in the writing of this manual. I also want to thank my husband for his continual support and for always being my best friend.

Susan Herring
Sonoma State University
January 2011

TABLE OF CONTENTS

PRELIMINARY CHAPTER
INTRODUCTION TO THE TI-83/84 PLUS

LESSON P.1   BASICS OF THE TI-83/84 PLUS

The TI-83/84 Plus is turned on by pressing the [ON] key located in the bottom left hand corner. It is turned off by pressing [2nd] OFF. Note that the TI-83/84 Plus has labeled keys and above the keys are symbols and commands printed in yellow and green. The yellow symbols and commands are accessed by pressing the yellow [2nd] key followed by the desired key, and the green letters are accessed by pressing the green [ALPHA] key followed by the desired key.

The screen of the TI-83/84 Plus will lighten as the batteries are used. To darken the screen, press [2nd] then press and hold [▲] until the screen is the desired darkness. Press [2nd], then press and hold [▼] to lighten the screen.

When the TI-83/84 Plus has low batteries, a message will appear on the home screen stating this. The TI-83/84 Plus has two different types of batteries. You will probably only need to change the AAA batteries. The backup battery is a lithium battery designed to last years. When changing the batteries, programs and lists stored in memory may be lost. NEVER remove both the lithium battery and the AAA batteries at the same time. When changing AAA batteries, do not remove all four at the same time, but remove and replace each battery individually until all four have been changed. This will minimize the chance of loosing stored programs and lists. When possible, it is best to backup programs and lists using a computer or another calculator before changing batteries.

The TI-83/84 Plus also has a catalog menu. This menu lists all the built-in functions and commands on the calculator and allows you to use them. However, it does not tell you where to locate the function or command. If you can not remember where a certain function or command is located, press [2nd] followed by [0] to access the catalog. You may arrow through the list using the arrow keys or jump to a section of the list by pressing the key associated with the letter at the beginning of the command you want.

LESSON P.2   USING THE TI-83/84 PLUS

Storing Data
Roughly speaking, data are the information collected in a statistical study and a data set consists of a particular collection of data. Each data set is stored in a list. There are several ways to enter data into the TI-83/84 Plus. They are entering the data directly, transferring the data from one calculator to another, and transferring the data from a disk to the calculator via a computer and the TI-graph link. The method you choose will depend on what is available for your use. We will demonstrate each of these methods in the following examples.

Example P.1   Typing in a List
In 1908, W. S. Gosset published a pioneering paper entitled "The Probable Error of a Mean" (*Biometrika*, Vol. 6, pp.1-25). Table P.1 displays one of several data sets discussed in the paper. The data gives the additional sleep, in hours, obtained by 10 patients using laevohysocyamine hydrobromide. Store these data in a list named HOUR by manually entering them into the TI-83/84 Plus.

Table P.1	1.9	.8	1.1	0.1
	-0.1	4.4	5.5	1.6
	4.6	3.4		

Solution: There are two ways a list can be entered and given a name. One is to enter the data into a generic list and then name it. The other is to enter the data into a named list. We will demonstrate both beginning with the generic list.

The TI-83/84 Plus has 6 lists numbered 1 through 6 built into it. Other lists may be stored and named (using names up to 5 characters long) as long as there is memory available.

1. From the home screen, press [STAT]. Your screen should appear as in Figure P.1. You have now entered the statistics editor menu.

2. Press [1] or press [ENTER] to enter the stat editor. A screen similar to Figure P.2 will appear.

Figure P.1

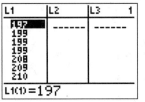

Figure P.2

If there is data listed under List 1, follow steps 3 and 4, otherwise skip to step 5.

3. To remove old data, before entering new data, arrow up until the name of the list is highlighted. See Figure P.3.

4. Press [CLEAR] and then arrow back down. Note: The list will not clear until the down arrow key is pressed. List 1 should now be clear as in Figure P.4

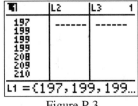

Figure P.3

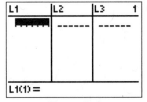

Figure P.4

5. Begin to enter your data by entering the number followed by pressing the down arrow key or [ENTER] after each entry. Note: The TI-83/84 Plus distinguishes between negative signs and subtraction. For negative signs be sure to use the gray [(-)] key located next to the [ENTER] key.

6. Be sure to check your entries to make sure all the numbers are correct. If an entry is wrong, use the arrow keys to highlight the entry and then type in the correct number pressing [ENTER] when you are done.

We have stored the data in List 1. The data will remain in the calculator even after the calculator has been turned off unless we clear the list or reset the calculator memory.

We will now resave this list as a list named HOUR.

7. From the home screen, press [2nd] L1 to indicate you wish to use List 1, then press [STO▸] [2nd] A-LOCK H O U R [ENTER]. Note: The [2nd] A-LOCK keys put the calculator into ALPHA mode. Look for the letters printed in green above the keys. See Figure P.5.

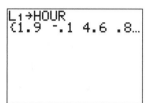

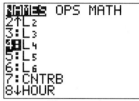

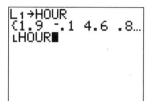

| Figure P.5 | Figure P.6 | Figure P.7 |

8. To access this list, press [2nd] LIST.   A list of all named lists will appear. See Figure P.6.

9. Press the number of the list, or arrow down to the name of the list you want and press [ENTER].   The name of the list preceded by an L will appear on your screen. See Figure P.7

10. Press [ENTER] and the list will be displayed.

11. To access the list within the stat editor, enter the editor by pressing [STAT] and press [ENTER].

12. Arrow up until the cursor highlights one of the lists, (here we will use List 2) and press [2nd] INS. A blank list with no name will appear. See Figure P.8.

13. Either spell out the name of the list you want using the green letters on the keyboard or press [2nd] LIST and press the number of the list you want. See Figure P.9. Press [ENTER] and the list will be displayed. See Figure P.10.

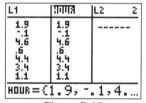

| Figure P.8 | Figure P.9 | Figure P.10 |

To remove a list from the stat editor, arrow up to the name of the list and press [DEL].

To return the stat editor to its factory setting of Lists 1-6 use the following steps.

1. Press [STAT] followed by [5] or arrow down to **5:SetUpEditor** and press [ENTER]. SetUp Editor will appear on your home screen.

2. Press [ENTER] and the calculator will respond with Done. If you reenter the stat editor, the lists 1 through 6 will be restored in order.

To type in a named list, the steps are similar.

1. Follow steps 1-2 above.

2. Arrow up until the cursor highlights one of the lists, (here we will use List 1) then arrow over until you have passed the last named list. A blank list with no name will appear. See Figure P.11.

3. Name the list using the letters printed in green above the keys. Here we will use the name HOUR. Spell out HOUR by pressing the keys with the appropriate green letter above them. (Note the calculator

automatically enters Alpha mode.) *The list name can not be more than 5 letters/numbers long and must start with a letter.* See Figure P.12.

4.   Begin to enter your data by entering the number followed by pressing the down arrow key or ENTER after each entry. Note: The TI-83/84 Plus distinguishes between negative signs and subtraction. For negative signs be sure to use the gray (-) key located next to the ENTER key.

5. Be sure to check your entries to make sure all the numbers are correct. If an entry is wrong, use the arrow keys to highlight the entry and then type in the correct number pressing ENTER when you are done. See Figure P.13.

6. Exit the stat list editor by pressing 2nd QUIT.

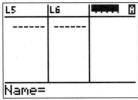

Figure P.11

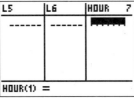

Figure P.12

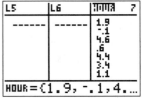

Figure P.13

Accessing a Named List

To access a named list on the home screen or in another menu press 2nd LIST and then select the list name by pressing its number or by arrowing down to the list and pressing ENTER.

LESSON P.3 SORTING DATA IN A LIST

Once your data is stored in a list, there are several ways you can manipulate the data. This includes sorting the data in ascending or descending order, and doing arithmetic on entire lists. The following examples illustrate sorting lists.

Example P.2   Use the TI-83/84 Plus to sort in ascending order the additional sleep, in hours, obtained by 10 patients for the data considered in Example P.1.

Solution: We assume the data is contained in List 1.

1. From the home screen, press STAT. Your screen will appear as in Figure P.14.

2. Press 2 or arrow down to **2:SortA(** and press ENTER. **SortA(** will appear on your home screen.

3. Enter the name of the list you wish to sort, here List 1, by pressing 2nd L1 ENTER. The calculator will respond with Done as in Figure P.15.

4. Reenter the Stat Editor by pressing 2nd ENTER. Your data will be sorted and appear as in Figure P.16.

Figure P.14

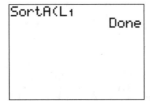

Figure P.15

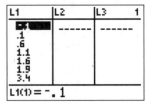

Figure P.16

The TI-83/84 Plus will sort in descending order if you use **3:SortD(** in place of **2:SortA(** in the above instructions. There may be times when you wish to sort more than one list according to the data in one of the lists. Example P.3 demonstrates this.

Example P.3   Anorexia nervosa is a serious eating disorder, found particularly among young women. The data in Table P.2 provide the weights, in pounds, of 17 anorexic young women before and after receiving a family-therapy treatment for anorexia nervosa. [SOURCE: Hand et al. (ed.) *A Handbook of Small Data Sets*, London:Chapman & Hall, 1994. Raw data from B. Everitt].

Table P.2

Before	After	Before	After	Before	After
83.3	94.3	76.9	76.8	82.1	95.5
86.0	91.5	94.2	101.6	77.6	90.7
82.5	91.9	73.4	94.9	83.5	92.5
86.7	100.3	80.5	75.2	89.9	93.8
79.6	76.7	81.6	77.8	86.0	91.7
87.3	98.0	83.8	95.2		

Use the TI-83/84 Plus to sort in descending order according to the before weight data the anorexic data on weights before and after treatment considered.

Solution: Begin by storing the before and after weight data in separate lists using one of the methods described above. We will assume that the before data has been stored in L_1 and the after data in L_2.

1. From the home screen, press STAT. Your screen will appear as in Figure P.17.

2. Press 3 or arrow down to **3:SortD(** and press ENTER. **SortD(** will appear on your home screen.

3. Enter the name of the list you wish to sort in descending order, here List 1, by pressing 2nd L1, followed by the lists you want sorted according to that list, here List 2, by pressing , 2nd L2 ENTER. The calculator will respond with Done as in Figure P.18.

4. Reenter the Stat Editor by pressing STAT ENTER. Your data will be sorted and appear as in Figure P.19.

Figure P.17

Figure P.18

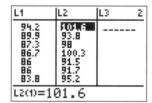
Figure P.19

You can use the same steps to sort several lists according to ascending order of one list by following the same steps but substituting **2:SortA(** for **3:SortD(**.

LESSON P.4   GRAPH WINDOWS

The TI-83/84 Plus is capable of doing several types of statistical graphs. We will discuss the graph window using the scatterplot.

Example P.4   Use the TI-83/84 Plus to obtain a plot of the after versus the before weights for the data considered in Example P.3.

Solution: Begin by storing the before and after data in separate lists using one of the methods described above. We will assume that the before data has been stored in L_1 and the after data in L_2. Note that the values to be plotted along the y-axis are listed first, then the values for the x-axis. Therefore we will plot the after weights along the y and the before weights along the x-axis.

Before attempting to graph any statistical plot, you must be sure that the function plot screen is cleared or that all function plots are turned off.

1. Press Y= to view the function plot screen. If the screen appears as in Figure P.20, then you are ready to plot your statistical graph, so proceed to step 5.

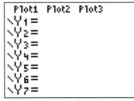   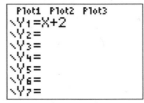

Figure P.20   Figure P.21

If you have a function in this window, such as in Figure P.21, then you may either clear the function or turn that function off.

2. To clear a function, arrow to the line where the function appears and press CLEAR. Repeat this for all functions you wish to clear. If you clear all functions, your screen will appear as in Figure P.20 above.

3. To turn off a function so that it does not graph, arrow to the line where the function appears and highlight the = of the function. Press ENTER. Arrow off the equal sign, and it will no longer be highlighted. See Figure P.22. This means the function is turned off. (Should you wish to turn it back on again, arrow to the = sign and press ENTER. The equal sign will once again be highlighted.)

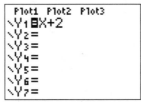

Figure P.22

4. Quit the function plot screen by pressing 2nd QUIT.

5. Enter the stat plot screen by pressing 2nd STAT PLOT. Your screen will appear similar to Figure P.23.

Figure P.23

The TI-83/84 Plus allows for up to 3 statistical plots to be graphed at one time. For this example we will use plot 1.

6. Enter the set-up screen for plot 1 by pressing [ENTER]. Your screen should appear similar to Figure P.24.

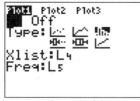   

Figure P.24   Figure P.25

7. Turn Plot 1 on by highlighting On and pressing [ENTER]. See Figure P.24.

8. Select the scatterplot by highlighting the scatterplot icon ⌇ (the first one in the Type list) and pressing [ENTER]. See Figure P.25.

9. The calculator must be told where the data is. We want to plot the before weights along the x-axis so arrow down to Xlist and press [2nd] L1 for L_1. Arrow down to Ylist and press [2nd] L2 to select L_2 to plot the after weights along the y-axis. See Figure P.25

10. Select the mark you would like to represent the points on the graph. For this example we will choose the box. Highlight the box and press [ENTER]. Your final screen should appear as in Figure P.25.

Now we must select the section of the graph that the TI-83/84 Plus will show you. Do this by entering the window screen. It is important to remember to adjust the window you are using each time you plot a new graph.

11. Press [WINDOW] to enter the window screen. Your screen should appear similar to Figure P.26.

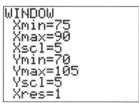

Figure P.26

12. Enter a Xmin value that is slightly lower than your lowest x-value. Here the x's are the before weights so we can enter a value of 75. Arrow down and enter a Xmax that is slightly above the highest x-value. Here we could use 90. Xscl sets how far apart your marks are on the x-axis, so pick a value based on your Xmin and Xmax. Here a suitable interval might be Xscl = 5.

13. Enter a Ymin value that is slightly lower than your lowest y-value. Here the y's are the after weights so we can enter a value of 70. Arrow down and enter a Ymax that is slightly above the highest y-value. Here we could use 105. Yscl sets how far apart your marks are on the y-axis, so pick a value based on your Ymin and Ymax. Here a suitable interval might be Yscl = 5. Your final window should appear as in Figure P.27.

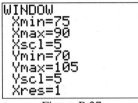

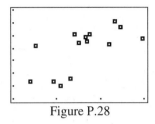

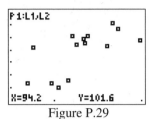

| Figure P.27 | Figure P.28 | Figure P.29 |

Note Xres is a setting for function graphing and we do not need to worry about it for statistical plots. Therefore, we will always leave it set at 1.

14. Press GRAPH to display the graph.   It should appear as in Figure P.28.

15. Because the TI-83/84 Plus does not display the values along the x and y-axes, the easiest way to identify a point is to use the Trace feature. Press TRACE. Your screen will appear as in Figure P.29. By using the left and right arrow keys, you can move the cursor from point to point and read off the values at the bottom of the screen. The up and down arrow keys allow you to switch between plots when plotting more than one graph on the screen at one time.

16. To clear the Trace data from the screen, press CLEAR. Press CLEAR again and the graph will clear and you will return to the home screen.

LESSON P.5   PROGRAMS FOR THE TI-83/84 PLUS

Several programs have been written for the TI-83/84 Plus to assist you with certain topics. These programs are contained in the WeissStatsCD.

There are three choices for entering a program into the calculator. The first is to type in the program. Refer to the calculator manual for what menus contain the commands used in the programs and how to enter a program.

The second is to download the program from a computer. To download a program from the WeissStatsCD, follow the directions for downloading a list from the WeissStatsCD. Open the Program directory/folder, and click on the program you wish to download.

The third option is to link a calculator with the program already entered to your calculator and transfer the program. The steps for this are similar to the steps for exchanging lists. In this case you will wish to select **3:Prgm** and then select the individual programs to exchange. Your teacher may choose to give you the programs this way.

All programs have been tested, but are not guaranteed.

To run a program, press PRGM and arrow down to the name of the program you wish to run. Press ENTER and the calculator will return to the home screen with the name of the program. Press ENTER again and the program will run.

CHAPTER 1
THE NATURE OF STATISTICS

LESSON 1.1 STATISTICS BASICS

Descriptive statistics consists of methods for organizing and summarizing information. Descriptive statistics includes the construction of graphs, charts, and tables and the calculation of various descriptive measures such as averages, measures of variation, and percentiles. We discuss descriptive statistics in detail in Chapters 2 and 3.

A **population** is the collection of all individuals or items under consideration in a statistical study. A **census** is when we obtain information for the entire population of interest. However, conducting a census may be time consuming, costly, impractical, or even impossible. Two methods other than a census for obtaining information are **sampling** and **experimentation.** In much of this book, we concentrate on sampling. When using sampling, we must choose the method for obtaining a sample from the population. Because the sample will be used to draw conclusions about the entire population, it should be a **representative sample**—that is, it should reflect as closely as possible the relevant characteristics of the population under consideration.

Inferential statistics consists of methods for drawing and measuring the reliability of conclusions about a population based on information obtained from a sample of the population. We will discuss inferential statistics in chapters 8 through 13.

LESSON 1.2 SIMPLE RANDOM SAMPLING

Because it is usually not practical to obtain information about an entire population, we use samples to infer information about the population of interest. **Simple random sampling** is a sampling procedure for which each possible sample of a given size is equally likely to be the one obtained. A **simple random sample (SRS)** is a sample obtained by simple random sampling.

The TI-83/84 Plus has several random number generators to aid us in this task. We will explain one of those random number generators here.

Example 1.8   During one semester, Professor Hassett wanted to sample the attitudes of the students taking college algebra at his school. He decided to interview 15 of the 728 students enrolled in the course. Since Professor Hassett had a registration list on which the 728 students were numbered 1-728, he could obtain a simple random sample of 15 students by randomly selecting 15 numbers between 1 and 728. Use the TI-83/84 Plus to obtain this simple random sample.

Solution: The TI-83/84 Plus uses a program to generate its random numbers and that program starts at the same number for every new calculator. We will first seed the random number generator. This changes the starting point. You only need to do this once, because the TI-83/84 Plus always uses its last random number as the new starting point.

To seed the generator, we will use the random number table located in the appendix of *Introductory Statistics* and *Elementary Statistics*. Close your eyes and put your finger down on the table. The 5 digit number you have chosen will be your new starting point.

1. On the home screen, enter the number you have selected. In this example we use 38634 as the new starting point.

2. Press STO▸ MATH and arrow over to the PRB menu. Your screen will appear as in Figure 1.1.

3. Press ENTER for the **rand** command.

4. Press [ENTER]. Your screen should appear similar to Figure 1.2.

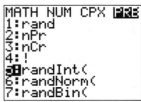

Figure 1.1   Figure 1.2   Figure 1.3

The TI-83/84 Plus has several random number generators used for different purposes. The first command is **rand** and it is used for generating numbers between 0 and 1. The second command is **randint** and it is used to generate integer values. We will use **randint** for this example.

5. Press [MATH] and arrow over to the PRB menu. Your screen should appear as in Figure 1.3.

6. Press [5] or arrow down to **5:randint(** and press [ENTER].

7. The format of this command is **randint(** lowerbound, upperbound, numsimulations). Here our lowest number is 1, our highest is 728 and we want 15 simulations. Therefore our command is **randint(1, 728, 15).** Because viewing will be easier in the list editor, we will store these values in L_1 by pressing [STO▸] [2nd] L1 [ENTER]. Your screen will appear similar to Figure 1.4. Of course, your random numbers will be different.

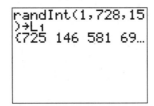
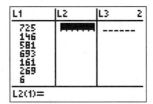

Figure 1.4   Figure 1.5

In Figure 1.5 we see the 15 randomly selected numbers are now stored in L_1. Professor Hassett could have used these 15 numbers as the registration numbers of the students to be interviewed.

LESSON 1.3 OTHER SAMPLING DESIGNS

Simple random sampling is the most natural and easily understood method of probability sampling because it corresponds to our intuitive notion of random selection by lot. However, simple random sampling does have drawbacks. For instance, it may fail to provide sufficient coverage when information about subpopulations is required and may be impractical when the members of the population are widely scattered geographically. In this section, we examine some commonly used sampling procedures that are often more appropriate than simple random sampling.

Remember, however, that the inferential procedures discussed in this book must be modified before they can be applied to data that are obtained by sampling procedures other than simple random sampling.

In **Systematic Random Sampling,** the population size is divided by the sample size and the result is rounded down to the nearest whole number, m. Then a random-number table or a similar device is used to obtain a number, k, between 1 and m. Finally a sample is selected from the population which are numbered k, k + m, k + 2m, …and so on.

Another sampling method is **cluster sampling,** which is particularly useful when the members of the population are widely scattered geographically. In cluster sampling, the population is divided into groups called clusters. Then a simple random sample of the clusters is obtained. Then all the members of the clusters are used as the sample.

Another sampling method, known as **stratified sampling,** is often more reliable than cluster sampling. In stratified sampling the population is first divided into subpopulations, called **strata,** and then sampling is done from each stratum. Ideally, the members of each stratum should be homogeneous relative to the characteristic under consideration.

Most large-scale surveys combine one or more of simple random sampling, systematic random sampling, cluster sampling, and stratified sampling. Such **multistage sampling** is used frequently by pollsters and government agencies.

LESSON 1.4 EXPERIMENTAL DESIGN

As we mentioned earlier, two methods for obtaining information, other than a census, are sampling and experimentation. In a designed experiment, the individuals or items on which the experiment is performed are called **experimental units.** When the experimental units are humans, the term **subject** is often used in place of experimental unit. The **response variable** is the characteristic of the experimental outcome that is to be measured or observed. A **Factor** is a variable whose effect on the response variable is of interest in the experiment. The possible values of a factor are called the **levels.** Each experimental condition is a **treatment**. For one-factor experiments, the treatments are the levels of the single factor. For multifactor experiments, each treatment is a combination of levels of the factors.

The following **principles of experimental design** enable a researcher to conclude that differences in the results of an experiment not reasonably attributable to chance are likely caused by the treatments.

> **Control:** Two or more treatments should be compared.
>
> **Randomization:** The experimental units should be randomly divided into groups to avoid unintentional selection bias in constituting the groups.
>
> **Replication:** A sufficient number of experimental units should be used to ensure that randomization creates groups that resemble each other closely and to increase the chances of detecting any differences among the treatments.

CHAPTER 2
ORGANIZING DATA

LESSON 2.1 VARIABLES AND DATA

A characteristic that varies from one person or thing to another is called a **variable.** Examples of variables for humans are height, weight, number of siblings, sex, marital status, and eye color. The first three of these variables yield numerical information and are examples of quantitative variables; the last three yield nonnumerical information and are examples of qualitative variables, also called categorical variables.

Some of the statistical procedures that you will study are valid for only certain types of data. This is one reason why you must be able to classify data. The classifications we have discussed are sufficient for most applications, even though statisticians sometimes use additional classifications.

LESSON 2.2 ORGANIZING QUALITATIVE DATA

Statistics is about organizing, summarizing, analyzing, and interpreting data. When the data is qualitative (also called categorical data) we can organize the data by forming frequency distributions, and relative frequency distributions. Graphs such as **pie charts** and **bar charts** can be used to describe the data. At the time of this writing, the TI-83/84 Plus does not have built in programs for performing the aforementioned tasks.

LESSON 2.3 ORGANIZING QUANTITATIVE DATA

It is said that a picture is worth a thousand words. Describing the distribution of data can be accomplished much more succinctly with pictures than with table. When the data is quantitative (also called numerical data) we can organize the data by constructing frequency distributions and relative frequency distribution. We can also graph the data using histograms, and dotplots.

Constructing Frequency Distributions and Relative Frequency Distributions
For a first look at data, it is practical to begin by describing the distribution of values in a data set. Many statistical tools have been developed for this purpose. Common tools used for organizing and summarizing a large number of data values are **frequency distributions** and **relative frequency distributions.**

The TI-83/84 Plus does not tabulate data from a list. Therefore we cannot use it to group our data. However, we can use it to calculate relative frequencies from the frequencies for each interval class.

Example 2.13   Consider the grouped data in Table 2.1 for the number of days to maturity of some 40 short-term investments. Compute the relative frequencies for these classes.

Table 2.1	Days	30-<40	40-<50	50-<60	60-<70	70-<80	80-<90	90-<100
	Freq.	3	1	8	10	7	7	4

Solution: We assume the frequencies are in list 1. First we find the total number of observations by obtaining the sum of the frequencies.

1. From the home screen, press 2nd LIST.

2. Arrow over to Math and press 5 or arrow to Math and then arrow to **5:sum(** and press ENTER. The **sum(** command will appear on your home screen.

3. Enter the list where your frequencies are stored. Here we press 2nd L1 to enter L_1.

4. Close the parentheses and press ENTER .   The sum is displayed. See Figure 2.1.

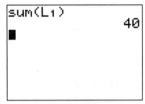

Figure 2.1

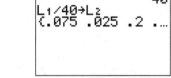

Figure 2.2

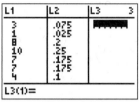

Figure 2.3

We find the relative frequencies by dividing the frequencies for each class by the total number of observations, which we have just found to be 40.

5. On the home screen press 2nd L1 ÷ 40 STO▶ 2nd L2 to compute the relative frequencies and store them in list 2. See Figure 2.2 for command and Figure 2.3 for results in the list editor.

2.3 Practice Problems

Problem 2.61 The Cheetah is the fastest land mammal on earth and is highly specialized to run down prey. According to the Cheetah Conservation of Southern Africa *Trade Environment Database*, the cheetah often exceeds speeds of 60 mph and has been clocked at speeds over 70 mph. Table 2.2 gives grouped data on the speeds, in miles per hour, over a quarter mile for 35 cheetahs. Compute the relative frequencies for these classes.

Table 2.2

Speed (mph)	Frequency
52-<54	2
54-<56	5
56-<58	6
58-<60	8
60-<62	7
62-<64	3
64-<66	2
66-<68	1
68-<70	0
70-<72	0
72-<74	0
74-<76	1

Problem 2.64 The grouped exam scores for the students in an introductory statistics class are given in Table 2.3. Compute the relative frequencies for these classes.

Table 2.3

Score	30-<40	40-<50	50-<60	60-<70	70-<80	80-<90	90-100
Freq.	3	1	8	10	7	7	4

Problem 2.66 Frustrated passengers, congested streets, time schedules, and air and noise pollution are just some of the physical and social pressures that lead many urban bus drivers to retire prematurely with disabilities such as coronary heart disease and stomach disorders. An intervention program designed by the Stockholm Transit District was implemented to improve the work conditions of the city's bus drivers. Improvements were evaluated by Evans et al. who collected physiological and psychological data for bus drivers who drove on the improved routes (intervention) and for drivers who were assigned the normal routes (control). Their findings were published in the article "Hassles on the Job: a Study of a Job Intervention With Urban Bus Drivers" (*Journal of Organizational Behavior* (Vol. 20, pp. 199-208). Table 2.4 contains the grouped data, based on the results of the study, for the heart rates, in beats per minute, of the control drivers. Compute the relative frequencies for these classes.

Table 2.4

Heart rate	50-<60	60-<70	70-<80	80-<90
Frequency	7	12	9	3

Histograms Using Cutpoint Grouping

A histogram depicts the classes along the horizontal axis and the frequencies on the vertical axis. The frequency of a class is represented by a vertical bar whose height is equal to the frequency of the class.

Example 2.15   The data for the number of days to maturity for 40 short-term investments is given in Table 2.5. Use the TI-83/84 Plus to construct a frequency histogram for these data based on the classes 30 - under 40, 40 - under 50,..., 90 - under 100.

Table 2.5

70	64	99	55	64	89	87	65
62	38	67	70	60	69	78	39
75	56	71	51	99	68	95	86
57	53	47	50	55	81	80	98
51	36	63	66	85	79	83	70

Solution: We will assume the maturity data are in L_1.

   1. Be sure that you have no graphs turned on in the [Y=] window.

   2. Press [2nd] STAT PLOT to begin setting up the STAT PLOT window. Your screen should look something like Figure 2.4.

   3. Choose which of the 3 plots you want to use and make sure that the others are turned off. Select a plot by pressing its number or by arrowing to it and pressing [ENTER]. For this example, we will use Plot 1.

   4. Using your arrow keys, highlight ON with the cursor and press [ENTER].

Figure 2.4

Figure 2.5

   5. Arrow down to Type and highlight ⌐⊓⌐, the histogram icon, by arrowing to the right and pressing [ENTER].

6. Arrow to Xlist and press [2nd] L1 to enter the list as L_1.

7. Arrow to frequency and set it as 1. Note: The TI-83/84 Plus will automatically go into alpha mode when it reaches the list and frequency entries. To return to normal mode, press [ALPHA]. Your screen should now appear as in Figure 2.5.

Now we must set the window. The window tells the TI-83/84 Plus which portion of the graph you wish to view. It is important to reset the window for every new graph.

8. Press [WINDOW].

9. For a histogram, the Xmin is the lowest value in your first class. Here that is 30.

10. The Xmax is the first value beyond your last class. For our example, that is 100.

11. The Xscl is your class width, 10 for this data set.

12. Set the Ymin at either -1 or 0. (If it is set at -1, viewing is easier when using the [TRACE] key. Therefore many people prefer to use -1.) The TI-83/84 Plus distinguishes between negatives and subtraction, so be sure to use the [(-)] key located next to [ENTER].

13. The Ymax is set just beyond your highest class frequency. If this is not known, estimate it. After graphing, use the [TRACE] key (explained below) to determine the highest class frequency and reset your window if necessary and graph again. Here, a Ymax of 11 will do.

14. Your Yscl should be set based on how far apart your Ymin and Ymax are. Here a Yscl of 1 is sufficient.

The Xres key sets the pixel resolution for function graphs only. For most of our graphs, its setting will not matter. Your window screen should appear as in Figure 2.6. Note: Although the TI-83/84 Plus has a ZoomStat feature it is generally not advisable to use it for graphing histograms.

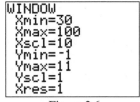
Figure 2.6

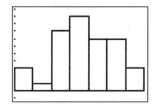
Figure 2.7

15. Once the plot and window screens are set, press [GRAPH]. The TI-83/84 Plus will display your histogram. See Figure 2.7.

Explanation of TRACE **key**

To identify the heights of the bars on your histogram, you can use the trace key. Press TRACE and notice that the TI-83/84 Plus lists the minimum and maximum for each class and the frequency as you arrow across the screen. See Figure 2.8.

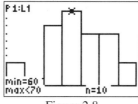

Figure 2.8

You may have stored the maturity data in a list named 'MATUR'. If so, the name of the list can be used when setting up the plot. Instead of typing L_1 at Xlist, type MATUR instead by pressing 2nd LIST, arrowing to the list MATUR and pressing ENTER.

2.3 Practice Problems (continued)

Problem 2.61 (continued)   The Cheetah is the fastest land mammal on earth and is highly specialized to run down prey. According to the Cheetah Conservation of Southern Africa *Trade Environment Database*, the cheetah often exceeds speeds of 60 mph and has been clocked at speeds over 70 mph. Table 2.6 gives grouped data on the speeds, in miles per hour, over a quarter mile for 35 cheetahs. Construct a frequency histogram for these data using 52 as the first cutpoint and classes of equal width 2.

Table 2.6

57.3	57.5	59.0	56.5	61.3	57.6	59.2
65.0	60.1	59.7	62.6	52.6	60.7	62.3
65.2	54.8	55.4	55.5	57.8	58.7	57.8
60.9	75.3	60.6	58.1	55.9	61.6	59.6
59.8	63.4	54.7	60.2	52.4	58.3	66.0

Problem 2.64 (continued)   The exam scores for the students in an introductory statistics class are given in Table 2.7. Construct a frequency histogram for these data using the classes 30-<40, 40-<50, 50-<60, … and so on.

Table 2.7

88	82	89	70	85
63	100	86	67	39
90	96	76	34	81
64	75	84	89	96

Problem 2.66 (continued)   Frustrated passengers, congested streets, time schedules, and air and noise pollution are just some of the physical and social pressures that lead many urban bus drivers to retire prematurely with disabilities such as coronary heart disease and stomach disorders. An intervention program designed by the Stockholm Transit District was implemented to improve the work conditions of the city's bus drivers. Improvements were evaluated by Evans et al. who collected physiological and psychological data for bus drivers who drove on the improved routes (intervention) and for drivers who were assigned the normal routes (control). Their findings were published in the article "Hassles on the Job: a Study of a Job Intervention With Urban Bus Drivers" (*Journal of Organizational Behavior* (Vol. 20, pp. 199-208). Table 2.8 contains the data, based on the results of the study, for the heart rates, in beats per minute, of the control drivers. Construct a frequency histogram for this data based on the classes 50-<60, 60-<70...

Table 2.8

74	52	67	63	77	57	80	77
53	76	54	73	54	60	77	63
60	68	64	66	71	66	55	71
84	63	73	59	68	64	82	

Histograms for Single-Value Grouped Data

When dealing with classes based on a single value, the bar for each class is centered over the only possible value for that class. Because we need the class value as well as the class frequency, this involves using two lists and changing the window values setup.

Frequency Histograms

Example 2.12   Consider the data on the number of TVs in each of 50 households given in Table 2.9 below. Construct a frequency histogram.

Table 2.9

No. of TVs	0	1	2	3	4	5	6
Frequency	1	16	14	12	3	2	2
Rel. Freq.	0.020	0.320	0.280	0.240	0.060	0.040	0.040

Solution: We assume that the data for the number of TVs is in List 1 and the frequencies in List 2. As in the previous example, we set up the STAT PLOT.

1. Press [2nd] STAT PLOT.

2. Choose which of the 3 plots you want to use and make sure that the others are turned off. For this example use Plot 1. Press [ENTER] to enter the setup screen for plot 1.

3. Using your arrow keys, highlight ON with the cursor and press [ENTER].

4. Arrow down to Type and highlight ⊞, the histogram icon and press [ENTER].

5. Arrow to Xlist and press [2nd] L1 to enter the list as L_1.

6. Arrow to Freq: and press [2nd] L2 to enter the list as L_2.

Now we must set the window.

7. For the bar to center over the class value, it is necessary to enter a Xmin that is halfway below the first class value and the next lower number. In this case, that would be - 0.5.

8. The Xmax should be halfway between the last class value and the next number above it. In this case, that would be 6.5.

9. Enter the Xscl as 1.

10. The Ymin, Ymax, Yscl should be determined based on the data as before. In this case, appropriate values would be Ymin = -1, Ymax = 20 and Yscl = 2. Press GRAPH. The graph should appear as in Figure 2.9.

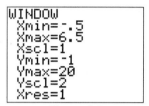

Figure 2.9

Relative Frequency Histogram

Example 2.12 (continued) Use the TVs per household data from Example 2.12 to graph a relative-frequency histogram.

Solution: We will assume the number of children are in List 1 and the relative frequencies are in List 3.

1. To set up the STAT PLOT window, proceed as in Example 2.3 steps 1-4 but set the Freq: as L_3.

2. For the window setting, the Xmin, Xmax and Xscl will remain as before, but the Y values must be set for relative frequencies. Because relative frequencies must always be between 0 and 1 inclusive, a window with settings of Ymin = -0.1 Ymax = 1.1 and Yscl = 0.1 will always show the entire graph. However, it will probably be somewhat condensed. See Figure 2.10. To spread out the graph a bit, pick a Ymax that is just slightly higher than your highest relative frequency. In this case a value like Ymax = 0.4 will work. See Figure 2.11. Note: Although the TI-83/84 Plus has a ZoomStat feature, it is generally not advisable to use it for graphing relative-frequency histograms.

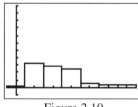

Figure 2.10

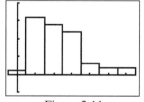

Figure 2.11

Notice the basic shape of both histograms is the same but the scaling of the second graph looks more like the original frequency histogram.

For data where the values are not grouped into single-value groups, a relative-frequency histogram can be graphed by entering the midpoints of the class into one list with the corresponding relative frequency entered into another list. Then treat the data as if it were single-value grouped data and construct the histogram.

2.3 Practice Problems (continued)

Problem 2.52 Professor Weiss asked his introductory statistics students to state how many siblings they have. Table 2.10 contains the results. Construct both a frequency histogram and a relative frequency histogram of this data.

Table 2.10	Siblings	0	1	2	3	4
	Frequency	8	17	11	3	1
	Rel. Freq.	0.200	0.425	0.275	0.075	0.025

Problem 2.53 The U.S. Bureau of the Census conducts nationwide surveys on characteristics of U.S. households and publishes the results in *Current Population Reports*. Table 2.11 contains data on the number of people per household for a sample of 40 households. Construct both a frequency histogram and a relative frequency histogram of this data.

Table 2.11	No. of people	1	2	3	4	5	6	7
	Freq.	7	13	9	5	4	1	1
	Rel. Freq.	0.175	0.325	0.225	0.125	0.1	0.25	0.25

Problem 2.76 The number of pups borne in a lifetime by 80 Great White Shark females is given in Table 2.12. Construct both a frequency histogram and a relative frequency histogram of this data.

Table 2.12	Pups	Freq.	Rel. Freq.
	3	2	0.03
	4	5	0.06
	5	10	0.12
	6	11	0.14
	7	17	0.21
	8	17	0.21
	9	11	0.14
	10	4	0.05
	11	2	0.03
	12	1	0.01

Review
Problem #21

The Prescott National Bank has six tellers available to serve customers. Data obtained from 25 spot checks on the number of busy tellers observed is contained in Table 2.13. Construct both a frequency histogram and a relative frequency histogram of this data.

Table 2.13	No. of tellers	0	1	2	3	4	5	6
	Frequency	1	2	2	4	5	7	4
	Rel. Freq.	0.04	0.08	0.08	0.16	0.2	0.28	0.16

Dotplots
A dotplot is constructed by drawing a horizontal axis that displays the values in the data set, and then placing a dot above each value corresponding to a particular piece of data. Although the TI-83/84 Plus does not have a built in dotplot, it is fairly easy to construct one.

Example 2.16   A man was interested in adding a new DVD player to his home theater system. He decided to use the Internet to shop and went to Price Watch U.S.A. There he found 16 quotes on different brands and styles of DVD players. The prices, in dollars, are depicted in Table 2.14. Construct a dotplot for these data.

Table 2.14	210	212	209	199	212	215	199	210
	197	199	224	219	208	219	214	219

Solution:

1. Enter the data into L_1 and sort the data.

2. Return to the stat editor and place your cursor at the beginning of L_2.

3. For each piece of data, record the cumulative frequency for that value in L_2. For example, the first data value of 197 has a cumulative frequency of 1. The first 199 would have a value of 1 but the second 199 would have a value of 2. The third 199 would have a value of 3 and the first value of 208 would have a value of 1, and so forth.

4. Go to the STAT PLOT window and enter Plot 1 by pressing [2nd] STAT PLOT [ENTER].

5. Highlight ON and press [ENTER].

6. Arrow down to Type, highlight ⋰, the scatterplot icon, and press [ENTER].

7. Arrow to Xlist and press [2nd] L1 to enter the list as L_1.

8. Arrow to Ylist and press [2nd] L2 to enter the list as L_2.

9. Choose the mark you would like to use by highlighting it and pressing [ENTER]. For this example, we chose the + mark. Note: It is usually better to use the box or the + sign to prevent confusion with the tic marks representing your y-axis. See Figure 2.12.

Figure 2.12

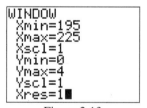

Figure 2.13

10. Set the window. As before, choose a Xmin that is slightly below your smallest x-value (here Xmin = 195) and a Xmax slightly above your largest x-value (here Xmax = 225). The Xscl should always be one. Make the Ymin = 0 and the Ymax slightly above your highest frequency (here Ymax = 4). Make the Yscl =1. See Figure 2.13. Press [GRAPH]. See Figure 2.14.

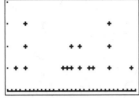

Figure 2.14

Figure 2.15

You may also use the ZoomStat feature on scatterplots rather than set the window. This is located under the ZOOM menu and is selected by pressing ZOOM and arrowing down to number 9 and pressing ENTER or by pressing ZOOM 9. This will automatically set the window and then graph. See Figure 2.15. The graph is basically the same but the window is slightly different.

2.3 Practice Problems (continued)

Problem 2.53 (continued) The U.S. Bureau of the Census conducts nationwide surveys on characteristics of U.S. households and publishes the results in *Current Population Reports*. Table 2.15 contains data on the number of people per household for a sample of 40 households. Construct a dotplot of this data.

Table 2.15

2	5	2	1	1	2	3	4
1	4	4	2	1	4	3	3
7	1	2	2	3	4	2	2
6	5	2	5	1	3	2	5
2	1	3	3	2	2	3	3

Problem 2.65 The Motor Vehicle Manufacturers Association of the United States publishes information on the ages of cars and trucks currently in use in *Motor Vehicle Facts and Figures*. A sample of 37 trucks provided the ages, in years, displayed in Table 2.16. Construct a dotplot for the ages.

Table 2.16

8	12	14	16	15	5	11	13
4	12	12	15	12	3	10	9
11	3	18	4	9	11	17	
7	4	12	12	8	9	10	
9	9	1	7	6	9	7	

Review Problem #21 (continued) The Prescott National Bank has six tellers available to serve customers. Data obtained from 25 spot checks on the number of busy tellers observed is contained in Table 2.17. Construct a dotplot for the data.

Table 2.17

6	5	4	1	5
6	1	5	5	5
3	5	2	4	3
4	5	0	6	4
3	4	2	3	6

2.4 DISTRIBUTION SHAPES

The distribution of a data set is a table, graph, or formula that provides the values of the observations and how often they occur. The distribution of a data set can be portrayed by frequency distributions, relative frequency distributions, histograms, stem-and-leaf diagrams, pie charts, and bar charts.

An important aspect of the distribution of a quantitative data set is its shape. To identify the shape of a distribution, the best approach usually is to use a smooth curve that approximates the overall shape.

When considering the shape of a distribution, you should also observe its number of peaks (highest points). A distribution is **unimodal** if it has one peak, **bimodal** if it has two peaks, and **multimodal** if it has three or more peaks.

When describing the shape of a distribution you should also consider whether the distribution is symmetric or skewed. If the distribution can be divided into two pieces that are mirror images of each other then we call the distribution symmetric. A unimodal distribution that is not symmetric, is either right-skewed or left-skewed.

LESSON 2.5 MISLEADING GRAPHS

Graphs and charts are frequently misleading, sometimes intentionally and sometimes inadvertently. Regardless of intent, we need to read and interpret graphs and charts with a great deal of care. You should also be careful to make graphs that clearly present the material.

CHAPTER 3
DESCRIPTIVE MEASURES

LESSON 3.1   MEASURES OF CENTER

Measures of central tendency are descriptive measures that indicate where the center or most typical value of a data set occurs. In this section, we will use the TI-83/84 Plus to find the two most important measures of center, the mean and the median.

The Mean
Recall the definition of the mean of a data set is the sum of the observations divided by the number of observations. It is also commonly called the average. There are two ways to find the mean of a list on a TI-83/84 Plus. The method we will use for finding the mean also provides several other descriptive measures.

The Median
Recall the definition of the median of a data set is the number that divides the bottom 50% of the data from the top 50%.

Example 3.1   A mathematician spent one summer working for a small mathematical consulting firm. The firm employed a few senior consultants, who made between $800 and $1050 per week; a few junior consultants, who made between $400 and $450 per week; and several clerical workers, who made $300 per week. Due to more work, more employees were employed the first half of the summer than the second half. Table 3.1 displays the weekly earnings for the first half of the summer. Find the mean and median.

Table 3.1	300	300	300	940	300	300	400
	300	400	450	800	450	1050	

Solution: Begin by entering the data from Table 3.1 into List 1.

1. Press $\boxed{\text{STAT}}$.

2. Arrow over to CALC.

3. Press $\boxed{1}$ or $\boxed{\text{ENTER}}$. **1 - Var Stats** will appear on your home screen. See Figure 3.1.

4. Enter the list where the data are stored, here List 1, $\boxed{\text{2nd}}$ L1, and press $\boxed{\text{ENTER}}$. See Figure 3.2.

Figure 3.1

Figure 3.2

5. This screen yields the mean as well as several other statistics. See Figure 3.3. Notice in Figure 3.3 there is an arrow on the bottom left corner next to n=13. By scrolling down, we see there are more statistics calculated. See Figure 3.4.

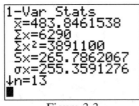   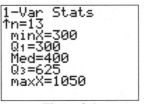

Figure 3.3   Figure 3.4

We see in Figure 3.3 that the mean is $483.85. We see in Figure 3.4 that the median is $400.

3.1 Practice Problems

Problem 3.15 In a study of the effects of radiation on amphibian embryos titled "Shedding Light on Ultraviolet Radiation and Amphibian Embryos" (*BioScience*, Vol. 53, No. 6, pp. 551-561), L. Licht recorded the time it took for a sample of seven different species of frogs and toads eggs to hatch. Table 3.2 shows the times to hatch, in days. Compute the sample mean and median.

Table 3.2	6	7	11	6	5	5	11

Problem 3.16 A recent article by D. Schaefer et al. (*Journal of Tropical Ecology*, Vol. 16, pp. 189-207) reported on a long term study of the effects of hurricanes on tropical streams of the Luquillo Experimental Forest in Puerto Rico. The study shows that Hurricane Hugo had a significant impact on stream water chemistry. Table 3.3 contains the data for a sample of 10 ammonia fluxes in the first year after Hugo. Data are in kilograms per hectare per year. Compute the sample mean and median.

Table 3.3	96	66	147	147	175	116	57	154	88	154

Problem 3.17 Each year, tornadoes that touch down are recorded by the Storm Prediction Center and published in *Monthly Tornado Statistics*. Table 3.4 gives the number of tornadoes that touched down in the United States during each month of the year 2002. [SOURCE: National Oceanic and Atmospheric Association.] Compute the sample mean and median.

Table 3.4	3	2	47	118	204	97
	68	86	62	57	98	99

Problem 3.18 In one Winter Olympics, Michelle Kwan competed in the Short Program ladies singles event. She received a score from nine judges ranging from 1 (poor) to 6 (perfect). Table 3.5 provides the marks that the judges gave her on technical merit, found in an article by S. Berry (Chance, Vol. 15, No. 2, pp. 14-18). Compute the sample mean and median.

Table 3.5	5.8	5.7	5.9	5.7	5.5	5.7	5.7	5.7	5.6

The Mode
Recall the mode of a data set is the value that occurs most often in a data set.

The TI-83/84 Plus does not have a mode command. The easiest way to determine the mode of a data set using the TI-83/84 Plus is to sort the data in a list and then count the number of times each data value appears.

Summation Notation
We often need the sum of a set of values. We use $\sum x$ to denote the sum of the set of x-values. The TI-83/84 Plus can be used to sum up the observations in a data list.

Example 3.5 A student has exam scores for a class as follows: 88, 75, 95, and 100. Find the sum of this data using the TI-83/84 Plus.

Solution: Enter the exam scores into List 1.

    1. Press $\boxed{\text{STAT}}$.

    2. Arrow over to CALC.

    3. Press $\boxed{1}$ or $\boxed{\text{ENTER}}$. **1 - Var Stats** will appear on your home screen. See Figure 3.1.

    4. Enter the list where the data are stored, here List 1, $\boxed{\text{2nd}}$ L1, and press $\boxed{\text{ENTER}}$. See Figure 3.2.

Figures 3.5 and 3.6 show the summary statistics for the data. In Figure 3.5 we see $\sum x = 358$. Thus, the sum of the exam scores is 358.

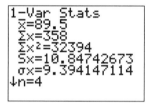

| Figure 3.5 | Figure 3.6 |

LESSON 3.2 MEASURES OF VARIATION

The Range

Recall the range of a data set is the difference between its maximum and minimum values.

The Sample Standard Deviation

Recall that the sample standard deviation of a data set is a measure of how far, on the average, the observations are from the mean. There are two formulas that can be used to compute the sample standard deviation.

$$s = \sqrt{\frac{\sum \left(x - \bar{x} \right)^2}{n-1}} \quad \text{and} \quad s = \sqrt{\frac{\sum x^2 - \left(\sum x \right)^2 / n}{n-1}}$$

In Lesson 3.2, we discuss how to compute the various pieces of these formulas individually. The TI-83/84 Plus will compute the sample standard deviation directly.

Examples 3.11-3.12: Let's look at the five starting players on each of two men's college basketball teams and their heights. Their heights in inches are given in Table 3.6. Find the range and standard deviation of the heights for Team I.

Table 3.6	Team I	72	73	76	76	78
	Team II	67	72	76	76	84

Solution: We assume the data for Team I is in List 1.

    1. Press $\boxed{\text{STAT}}$.

    2. Arrow over to CALC.

    3. Press $\boxed{1}$ or $\boxed{\text{ENTER}}$. **1 - Var Stats** will appear on your home screen. See Figure 3.1.

4. Enter the list where the data are stored, here List 1, [2nd] L1, and press [ENTER]. See Figure 3.2.

5. To find the range, we need to find the maximum and minimum value. In Figure 3.8 we see the maximum height is 78 inches and the minimum height is 72 inches. Therefore, the range is 78 - 72 = 6 inches.

6. In Figure 3.7 we see the sample standard deviation is s= 2.449 rounded to 3 decimal places.

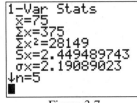

Figure 3.7   Figure 3.8

To obtain the sample variance of a data set, simply square the sample standard deviation. The easiest way to do this is by typing [2nd] ANS [x²] immediately after you have computed the sample standard deviation.

3.2 Practice Problems

Problem 3.71 Find the range and sample standard deviation for the times to hatch, in days, given in Problem 3.15.

Problem 3.72 Find the range and sample standard deviation for the ammonia fluxes given in Problem 3.16.

Problem 3.73 Find the range and standard deviation for the number of tornadoes given in Problem 3.17.

Problem 3.74 Find the range and standard deviation for the marks for technical merit given in Problem 3.18.

LESSON 3.3 THE FIVE NUMBER SUMMARY; BOXPLOTS

Recall that the five-number summary of a data set consists of the minimum, the quartiles, and the maximum, written in increasing order.

Example 3.17 Consider the A.C. Nielsen Company data on the viewing habits of 20 Americans. The weekly viewing times, in hours, are displayed in Table 3.7. Find the sample mean, sample standard deviation, and the five number summary of this data.

Table 3.7				
25	41	27	32	43
66	35	31	15	5
34	26	32	38	16
30	38	30	20	21

Solution: We assume that the data are in List 1.

1. Press [STAT].

2. Arrow over to CALC.

3. Press [1] or [ENTER]. **1 - Var Stats** will appear on your home screen.

4. Enter the list where the data are stored, here List 1, [2nd] L1, and press [ENTER]. See Figure 3.9. This screen yields the sample mean, sum of the data, sum of the squared data, the sample standard deviation, the population standard deviation, and the sample size. Arrow down to find, the minX, Q_1, Med, Q_3, and maxX are displayed. The last five values are the five number summary. See Figure 3.10.

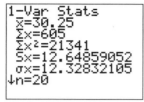

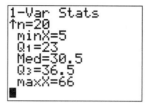

Figure 3.9   Figure 3.10

Note that different software programs compute the first and third quartiles differently. If the position number is not a whole number, the TI-83/84 Plus averages the two numbers in the closest positions.

Boxplots
Recall that the boxplot and modified boxplot are based on the five-number summary.

Example 3.19   Let's return to the A.C. Nielsen TV viewing data given in Table 3.7 of Example 3.17. We will assume that this data is in List 1. Construct a boxplot and a modified boxplot.

Solution: First, we will construct a boxplot.

1. Setup the Stat Plot by pressing [2nd] STAT PLOT.

2. Enter Plot 1 and turn it on by highlighting On and pressing [ENTER].

3. The boxplot icon is the fifth one under Type. Highlight ⊡⊢ and press [ENTER].

4. For Xlist enter L_1 and set the frequency as 1. See Figure 3.11.

5. Set your window by choosing Xmin slightly lower than your data's minimum, and Xmax slightly above your data's maximum. Choose an appropriate Xscl based on how far apart your max and min are. For our problem, we can use Xmin = 0, Xmax = 70, and Xscl = 5.

6. The Y values do not affect boxplots as the TI-83/84 Plus will always plot the first boxplot at the top of the screen, the second in the middle and the third on the bottom. However, it is generally advisable not to have the y-axis cutting through a boxplot. To prevent this, set Ymin = 0. You can also use ZoomStat to set the window for you.

A modified boxplot is set the same way, only the modified boxplot icon is the fourth icon under Type, ·⊡··.

7. Enter Plot 2 and set it up to graph a modified boxplot for this data.

8. Choose the mark to use for any outliers. Here the box is chosen.

9. When you are finished setting the plot window (see Figure 3.12) press [GRAPH]. Both graphs are plotted in Figure 3.13.

Figure 3.11

Figure 3.12

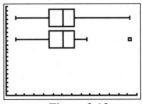

Figure 3.13

You may use TRACE to find the five number summary and identify any outliers.

3.3 Practice Problems

Problem 3.123   The U.S. National Center for Health Statistics compiles data on the length of stay by patients in short-term hospitals and publishes its findings in *Vital and Health Statistics*. A random sample of 21 patients yielded the data on length of stay, in days given in Table 3.8. Find the five number summary. Construct a (modified) boxplot.

Table 3.8	4	4	12	18	9	6	12
	3	6	15	7	3	55	1
	10	13	5	7	1	23	9

Problem 3.124   The U.S. Federal Highway Administration conducts studies on motor vehicle travel by type of vehicle. Results are published annually in *Highway Statistics*. A sample of 15 cars yields the data in Table 3.9 on number of miles driven, in thousands, for last year. Find the five number summary. Construct a (modified) boxplot.

Table 3.9	13.2	13.3	11.9	15.7	11.3
	12.2	16.7	10.7	3.3	13.6
	14.8	9.6	11.6	8.7	15.0

Problem 3.125   An article by D. Schaefer et al. (*Journal of Tropical Ecology*, Vol. 16, pp. 189–207) reported on a long-term study of the effects of hurricanes on tropical streams of the Luquillo Experimental Forest in Puerto Rico. The study shows that Hurricane Hugo had a significant impact on stream water chemistry. Table 3.10 shows a sample of 10 ammonia fluxes in the first year after Hugo. Data are in kilograms per hectare per year. Find the five number summary. Construct a (modified) boxplot.

Table 3.10	96	66	147	147	175
	116	57	154	88	154

LESSON 3.4   DESCRIPTIVE MEASURES FOR POPULATIONS; USE OF SAMPLES

When possible, it is preferable to use descriptive measures based on the entire population. However, it is rare that we have access to the entire population. Nevertheless, we will examine the formulas for these measures.

Population Mean
The formula for the population mean is the same as for the sample mean, although the symbol is the Greek letter μ.

$$\mu = \frac{\sum x}{N} \text{ where N is the population size}$$

Because the formulas are the same, the TI-83/84 Plus does not distinguish between the sample mean and the population mean. The computations are done the same way.

Population Standard Deviation

The formula for the population standard deviation is different than the formula for the sample standard deviation. The symbol used for population standard deviation is the Greek letter σ. Like the sample standard deviation, there are two formulas for computing the population standard deviation. They are

$$\sigma = \sqrt{\frac{\sum (x - \mu)^2}{N}} = \sqrt{\frac{\sum x^2}{N} - \mu^2}$$

Example 3.22   From the Universal Sports Web site, we obtained data for the players on the 2008 U.S. women's Olympic soccer team, as shown in Table 3.11. Heights are given in centimeters (cm) and weights in kilogram (kg). Find the population mean and the population standard deviation of the weight of these soccer players.

Table 3.11

Name	Weight (kg)
Barnhart, Nicole	73
Boxx, Shannon	67
Buehler. Rachel	68
Chalupny, Lori	59
Chaney, Lauren	72
Cox, Stephanie	59
Heath, Tobin	59
Hucles, Angela	64
Kai, Natasha	65
Lloyd, Carli	65
Markgraf, Kate	61
Mitts, Heather	55
O'Reilly, Heather	59
Rampone, Christie	61
Rodriguez, Amy	59
Solo, Hope	64
Tarpley, Lindsay	59
Wagner, Aly	57

Solution: We will assume this data is stored in L_1.

1. From the home screen, press [STAT].

2. Arrow over to CALC.

3. Press [1] or [ENTER]. 1 - Var Stats will appear on your home screen.

4. Type in the list where the data are stored, here List 1, [2nd] L1, and press [ENTER]. The descriptive measures will appear on your screen. The population mean is the same as the sample mean, but σ is listed separately from **s**. See Figure 3.14.

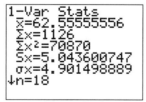

Figure 3.14

We see the population mean is 62.556 rounded to 3 decimal places and that the population standard deviation is 4.901 rounded to 3 decimal places. Note that the population standard deviation may differ from values given in the textbook. This difference is because the book may round the population mean whereas the calculator will always use the non-rounded mean.

To obtain the population variance, simply square the population standard deviation.

CHAPTER 4
PROBABILITY CONCEPTS

LESSON 4.1 PROBABILITY BASICS

In many experiments, all the outcomes are **equally likely**. When the outcomes are all equally likely, the probabilities are percentages or relative frequencies. Suppose that an experiment can result in N equally likely outcomes. Simple random sampling is an example of an experiment where all the outcomes are equally likely. Suppose that a specific event, A, consists of f of the N possible outcomes. The probability of the event, A, is the number of ways, f, that the event can occur, divided by the total number of possible outcomes. That is, the probability is the percentage of the N outcomes that are associated with the event A. In symbols,

$$P(A) = \frac{\#\ of\ outcomes\ in\ A}{Total\ \#\ of\ outcomes} = \frac{f}{N}$$

LESSON 4.2 EVENTS

The *sample space* for en experiment is the collection of all possible outcomes for the experiment. An *event* is a collection of outcomes for the experiment. An outcome occurs if and only if the outcome of the experiment is a member of the event. Two or more events are *mutually exclusive events* if no two of them have outcomes in common.

Example 4.2   The U.S. Bureau of the Census compiles data on family income and publishes its findings in *Current Population Reports*. Table 4.1 gives a frequency distribution of annual income for U.S. families.

Table 4.1

Income	Frequency (1000s)
Under $15,000	6,945
$15,000 - $24,999	7,765
$25,000 - $34,999	8,296
$35,000 - $49,999	11,301
$50,000 - $74,999	15,754
$75,000 - $99,999	10,471
$100,000 and over	16,886
	77,418

Find the probability that a randomly selected family has an annual income
   a.   between $50,000 and $74,999, inclusive.
   b.   between $15,000 and $49,999, inclusive.
   c.   under $25,000.

a. The probability a randomly selected family has an annual income between $50,000 and $74,999, inclusive is

$$\frac{15,754}{77,418} = 0.203$$

b. The number of families with annual incomes between $15,000 and $49,999, inclusive is 7,765 + 8,296 + 11,301 = 27,362. Therefore, the probability a randomly selected family has an annual income between $15,000 and $49,999, inclusive is

$$\frac{27,362}{77,418} = 0.353$$

c. The number of families with annual incomes under $25,000 is 6945 + 7765 = 14,710. Therefore, the probability a randomly selected family has an annual income between $15,000 and $49,999, inclusive is

$$\frac{14,710}{77,418} = 0.190$$

4.2 Practice Problems

Problem 4.16   As reported by the Federal Bureau of Investigation in *Crime in the United States*, the age distribution of murder victims between 20 and 59 years old is given in Table 4.2. A murder case in which the person murdered was between 20 and 59 years old is selected at random.

Table 4.2

Age	Frequency
20 – 24	2834
25 – 29	2262
30 – 34	1649
35 – 39	1257
40 – 44	1194
45 – 49	938
50 – 54	708
55 – 59	384

Find the probability that the randomly selected murder victim is

a.   between 40 and 44 years old, inclusive.
b.   at least 25 years old.
c.   between 45 and 59 years old, inclusive.
d.   is under 30 or over 54.

LESSON 4.3   SOME RULES OF PROBABILITY

If E is an event, then *P(E)* represents the probability that event E occurs, it is read as "the probability of the event E"

General Addition Rule: If A and B are any two events, then P(A or B) = P(A) + P(B) – P(A and B)

Complementation Rule: For any event E, P(E) = 1 – P(not E)

LESSON 4.4   CONTINGENCY TABLES; JOINT AND MARGINAL PROBABILITIES*

Data obtained by observing values of two variables of a population are called bivariate data, and a frequency distribution for bivariate data is called a contingency table or two-way table. Contingency tables become joint probability tables when their frequencies are replaced with probabilities. Example 4.15 illustrates how to do this.

Example 4.15   The *Arizona State University Statistical Summary* provides information on various characteristics of ASU faculty. Data on the variables age and rank of ASU faculty yielded the contingency table shown in Table 4.3. Create a joint probability distribution for this data.

Table 4.3

		Rank				
		Full professor R_1	Associate professor R_2	Assistant professor R_3	Instructor R_4	Total
Age	Under 30 A_1	2	3	57	6	68
	30-39 A_2	52	170	163	17	402
	40-49 A_3	156	125	61	6	348
	50-59 A_4	145	68	36	4	253
	60 & over A_5	75	15	3	0	93
	Total	430	381	320	33	1164

Solution: We will begin by entering this data into a matrix in the TI-83/84 Plus.

1. Press [2nd] MATRX, arrow over to EDIT, and press [ENTER].

2. Enter the number of rows, here 6, and press [ENTER].

3. Enter the number of columns, here 5, and press [ENTER].

4. Begin entering your data values going across the rows and pressing [ENTER] after each entry. Be sure you enter the totals, and that you press [ENTER] after the last entry. See Figure 4.1 for the final screen.

5. Exit the matrix edit mode by pressing [2nd] QUIT.

Now we must divide each of the entries in the matrix by 1164 to get our probabilities.

6. From the home screen, press [2nd] MATRX [ENTER] to get the name of the matrix .

7. Because matrix division is undefined we must multiply by the inverse of 1164 which is 1/1164. Therefore press [×] [(] [1] [÷] [1] [1] [6] [4] [)]. Your screen should appear as in Figure 4.2. Press [ENTER] and your screen should appear as in Figure 4.3.

8. Using the arrow keys, scroll across the screen to get the joint probability distribution as recorded in Table 4.4 where each probability is rounded to 3 decimal places.

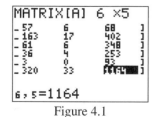

Figure 4.1

Figure 4.2

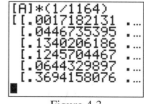

Figure 4.3

Table 4.4

Age		Rank: Full professor R_1	Associate professor R_2	Assistant professor R_3	Instructor R_4	$P(A_j)$
	Under 30 A_1	0.002	0.003	0.049	0.005	0.058
	30-39 A_2	0.045	0.146	0.140	0.015	0.345
	40-49 A_3	0.134	0.107	0.052	0.005	0.299
	50-59 A_4	0.125	0.058	0.031	0.003	0.217
	60 & over A_5	0.064	0.013	0.003	0.000	0.080
	$P(R_j)$	0.369	0.327	0.275	0.028	1.000

4.4 Practice Problems

Problem 4.97 The National Football League updates team rosters and posts them on their web site at www.nfl.com. Table 4.5 provides a cross-classification of players on the New England Patriots as of December 3, 2008, by weight and years of experience. Create a joint probability distribution table for this data.

Table 4.5

Weight (lb.)		Years of Experience: Rookie	1-5	6-10	10+	Total
	Under 200	3	4	1	0	8
	200-300	8	12	17	6	43
	Over 300	0	8	6	0	14
	Total	11	24	24	6	65

Problem 4.98 The U.S. Federal Highway Administration compiles information on motor vehicle use around the globe and publishes its findings in *Highway Statistics*. The number of motor vehicles in use in North American countries by country and type of vehicle is summarized in Table 4.6. Frequencies are in thousands. Create a joint probability distribution table for this data.

Table 4.6

Vehicle Type		Country: U.S.	Canada	Mexico	Total
	Automobiles	129,728	13,138	8,607	151,473
	Motorcycles	3,871	320	270	4,461
	Trucks	75,940	6,933	4,287	87,160
	Total	209,539	20,391	13,164	243,094

Problem 4.112   The *International Shark Attack File*, maintained by the American Elasmobranch Society and the Florida Museum of Natural History, is a compilation of all known shark attacks around the globe from the mid-1500s to the present. Table 4.7 provides a cross classification of worldwide reported shark attacks during the 1990s by country and lethality of attack. Create a joint probability distribution table for this data.

Table 4.7

Country	Lethality		
	Fatal	Non-fatal	Total
Australia	9	56	65
Brazil	12	21	33
South Africa	8	57	65
United States	5	244	249
Other	36	92	128
Total	70	470	540

LESSON 4.5   CONDITIONAL PROBABILITY*

The probability that an event occurs given that another event has occurred is called a **conditional probability**. The probability the event B occurs given the event A occurs is denoted by P(B|A), which is read the "the probability of B given A". We call A the given event.

$$P(B \mid A) = \frac{\text{\# of outcomes in A and B}}{\text{\# outcomes in A}}$$

Conditional probabilities are often used to analyze contingency tables.

Example 4.18   The data in Example 4.15 provides a contingency table showing the rank versus age of ASU faculty members. Suppose a faculty member is selected at random from all ASU faculty members. Obtain the conditional probability distribution for each age conditioning on rank. In this case, we are finding the conditional probability distribution for **age** (the row variable) conditioning on **rank** (the column variable).

In order to find the conditional probability distribution of age conditioning on rank, we need to find the probability of the faculty member being in each age group given the rank of the faculty member.

For example, for the rank of full professor, we will find the probability the full professor is in of each age group. Since there are 430 full professors, we will divide the number of full professors in each age group by the total number of full professors.

$$P(\text{Under 30}|\text{Full professor}) = \frac{2}{430} = 0.005$$

$$P(\text{30-39} \mid \text{Full professor}) = \frac{52}{430} = 0.121$$

$$P(\text{40-49} \mid \text{Full professor}) = \frac{156}{430} = 0.363$$

$$P(\text{60 \& over}| \text{Full professor}) = \frac{75}{430} = 0.174$$

In Table 4.8, theses probabilities are shown in the column for full professor. Continuing in the same way for the other ranks, we obtain the following conditional probability distribution for **age** (the row variable) conditioning on **rank** (the column variable). For example, if the randomly chosen faculty member were an Associate professor, the probability that he or she is between 40 and 49, inclusive is 0.328. All such probabilities can be read from the table.

Table 4.8   Rank

Age	Full professor R_1	Associate professor R_2	Assistant professor R_3	Instructor R_4	$P(A_j)$
Under 30 A_1	0.005	0.008	0.178	0.182	0.058
30-39 A_2	0.121	0.448	0.509	0.515	0.345
40-49 A_3	0.363	0.328	0.191	0.182	0.299
50-59 A_4	0.337	0.178	0.113	0.121	0.217
60 & over A_5	0.174	0.039	0.009	0.000	0.080
$P(R_j)$	1.000	1.000	1.000	1.000	1.000

Example 4.18 (continued):   The data in Example 4.15 provides a contingency table showing the rank versus age of ASU faculty members. Suppose a faculty member is selected at random from all ASU faculty members. Obtain the conditional probability distribution for each rank conditioning on age, that is, the conditional distribution for rank within age. We want the probability distribution of *rank* (the column variable) conditioned on *age* (the row variable).

In order to find the conditional probability distribution of rank conditioning on age, we need to find the probability of the faculty member being in each rank given the age group of the faculty member.

For example, for faculty under 30 years of age, we will find the probability the faculty member is in of each rank. Since there are 68 faculty under 30, we will divide the number of faculty under 30 in each rank by the total number of faculty under 30.

$$P(\text{Full professor}| \text{Under 30}) = \frac{2}{68} = 0.029$$

$$P(\text{Associate professor}| \text{Under 30}) = \frac{3}{68} = 0.044$$

$$P(\text{Assistant professor}| \text{Under 30}) = \frac{57}{68} = 0.838$$

$$P(\text{Instructor}| \text{Under 30}) = \frac{6}{68} = 0.088$$

In Table 4.9, theses probabilities are shown in the row for faculty under 30. Continuing in the same way for the other age groups, we obtain the following conditional probability distribution for **rank** (the column variable) conditioning on **age** (the row variable). This table gives us the conditional probability of the rank of a faculty member given the age of the faculty member. For example, if the randomly chosen faculty member were between 40 and 49, inclusive the probability that he or she is an Associate professor is 0.359. All such conditional probabilities can be read from the table.

Table 4.9

		Rank				
		Full professor R_1	Associate professor R_2	Assistant professor R_3	Instructor R_4	$P(A_j)$
Age	Under 30 A_1	0.029	0.044	0.838	0.088	1.000
	30-39 A_2	0.129	0.423	0.405	0.042	1.000
	40-49 A_3	0.448	0.359	0.175	0.017	1.000
	50-59 A_4	0.573	0.269	0.142	0.016	1.000
	60 & over A_5	0.806	0.161	0.032	0.000	1.000
	$P(R_j)$	0.369	0.327	0.275	0.028	1.000

4.5 Practice Problems

Problem 4.92 (continued) The U.S. Federal Highway Administration compiles information on motor vehicle use around the globe and publishes its findings in *Highway Statistics*. The number of motor vehicles in use in North American countries by country and type of vehicle is summarized in Table 4.10. Frequencies are in thousands. Create a joint probability distribution table for this data.

Table 4.10

	Country			
Vehicle Type	U.S.	Canada	Mexico	Total
Automobiles	129,728	13,138	8,607	151,473
Motorcycles	3,871	320	270	4,461
Trucks	75,940	6,933	4,287	87,160
Total	209,539	20,391	13,164	243,094

a. What is the probability that the randomly selected vehicle was an automobile from Canada?

b. What is the probability that the randomly selected vehicle was a truck from Mexico?

c. What is the conditional probability of a randomly chosen vehicle being from Canada given that it is an automobile?

d. Determine the conditional probability that a randomly chosen vehicle is Motorcycle given that it is from Mexico.

Problem 4.111 (continued) The National Football League updates team rosters and posts them on their web site at www.nfl.com. Table 4.11 provides a cross-classification of players on the New England Patriots as of December 3, 2008 by weight and years of experience.

Table 4.11

		Years of Experience				
		Rookie	1-5	6-10	10+	Total
Weight (lb.)	Under 200	3	4	1	0	8
	200-300	8	12	17	6	43
	Over 300	0	8	6	0	14
	Total	11	24	24	6	65

   a.   A player on the New England Patriots is selected at random. Find the probability that the player selected is a rookie.

   b.   A player on the New England Patriots is selected at random. Find the probability that the player selected weighs under 200 pounds.

   c.   A player on the New England Patriots is selected at random. Find the probability that the player selected is a rookie given he weighs under 200 pounds.

   d.   A player on the New England Patriots is selected at random. Find the probability that the player selected weighs under 200 pounds given he is a rookie.

Problem 4.112 (continued) The *International Shark Attack File*, maintained by the American Elasmobranch Society and the Florida Museum of Natural History, is a compilation of all known shark attacks around the globe from the mid-1500s to the present. Table 4.12 provides a cross classification of worldwide reported shark attacks during the 1990s by country and lethality of attack.

Table 4.12

		Lethality		
		Fatal	Non-fatal	Total
Country	Australia	9	56	65
	Brazil	12	21	33
	South Africa	8	57	65
	United States	5	244	249
	Other	36	92	128
	Total	70	470	540

   a.   Find the probability that an attack is in Brazil.

   b.   Find the probability an attack is in Brazil and fatal.

   c.   Find the probability of an attack being fatal given the attack was in Brazil.

LESSON 4.6 MULTIPLICATION RULE; INDEPENDENCE*

General Multiplication Rule: If A and B are any two events, then
$$P(A \text{ and } B) = P(A) \, P(B|A)$$

One of the most important concepts in probability is that of *statistical independence*, or more simply, *independence*. Event B is said to be *independent* of event A if $P(B|A) = P(B)$. If event B is independent of event A, then event A is also independent of event B. Thus, we can say that A and B are *independent events*.

The terms *mutually exclusive* and *independent* refer to different concepts. *Mutually exclusive events* are those that cannot occur simultaneously, whereas *independent events* are those for which the occurrence of some does not affect the probabilities of the others occurring. In fact, if two or more events are mutually exclusive, the occurrence of one precludes the occurrence of the others. Two or more (non-impossible) events cannot be both mutually exclusive and independent.

LESSON 4.7 BAYES'S RULE*

Bayes's Rule was developed by Thomas Bayes, an eighteenth-century clergyman. One of the primary uses of *Bayes's Rule* is to revise probabilities in accordance with newly acquired information. Such revised probabilities are actually conditional probabilities.

Bayes's Rule: Suppose the events A_1, A_2,....,A_k are mutually exclusive and exhaustive. Then for any event B,

$$P(A_i \mid B) = \frac{P(A_i)P(B \mid A_i)}{\sum_{j=1}^{k} P(A_j)P(B \mid A_j)}$$

where A_i can be any one of the events A_1, A_2,....,A_k.

LESSON 4.8 COUNTING RULES*

We often need to determine the number of ways something can happen. For example, the number of possible outcomes for an experiment, the number of ways an experiment can occur, and so on. Sometimes, we can list the possibilities and count them, but, usually, doing this is impractical. Therefore we need to develop techniques that do not rely on a direct listing for determining the number of ways something can happen. Such techniques are called *counting rules*.

Basic Counting Rule (BCR): Suppose that r actions are to be performed in a definite order. Further suppose that there are m_1 possibilities for the first action and that corresponding to each of these possibilities are m_2 possibilities for the second action, and so on. Then there are $m_1 \cdot m_2 \cdots m_r$ possibilities altogether for the r actions.

The product of the first *k* positive integers is called *k factorial* and is denoted *k!*. In symbols,

$$k! = k(k - 1) \cdots 2 \cdot 1.$$

We also define *0! = 1*.

The TI-83/84 Plus will compute a factorial for us. However, the maximum value it can compute is 70!.

Example 4.29   Use the TI-83/84 Plus to compute 5!

Solution:
1. Enter the number 5 on your home screen.

2. Press $\boxed{\text{MATH}}$ and arrow over to the PRB menu. See Figure 4.4.

3. Press $\boxed{4}$ or arrow down to **4:!** and press $\boxed{\text{ENTER}}$.

4. Press $\boxed{\text{ENTER}}$ and the result will be displayed. See Figure 4.5.

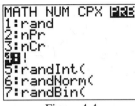
Figure 4.4

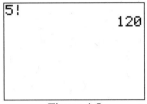
Figure 4.5

Note that the TI-83/84 Plus will automatically enter scientific notation as the numbers get larger.

A **permutation** of r objects from a collection of m objects is any ordered arrangement of r of the m objects. The number of possible permutations of r objects from a collection of *m* objects is given by the formula

$$_m P_r = \frac{m!}{(m-r)!}$$

The TI-83/84 Plus will compute permutations for us, although it uses $_n P_r$ rather than $_m P_r$.

Example 4.31   In an exacta wager at the race track, the bettor picks the two horses that he or she thinks will finish first and second, in a specified order. For a race with 12 entrants, determine the number of possible exacta wagers.

Solution: Because order matters, this is a permutation problem. We will need to find $_{12}P_2$.

1. On the home screen, enter the total number of objects, here 12.

2. Press ⌑MATH⌑ and arrow over to the PRB menu. See Figure 4.6.

3. Press ⌑2⌑ or arrow down to **2:nPr** and press ⌑ENTER⌑. The calculator will return to the home screen.

4. Enter the number of objects selected, here 2 and press ⌑ENTER⌑.   Your calculator will compute the answer and your screen will appear as in Figure 4.7.

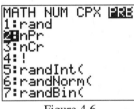
Figure 4.6

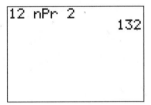
Figure 4.7

A **combination** of r objects from a collection of m objects is any unordered arrangement of r of the m objects. In other words, a combination is any subset of r objects from the collection of m objects. Note that order matters for permutations, but not in combinations. The number of possible combinations of *r* objects from a collection of *m* objects is given by the formula

$$_m C_r = \frac{m!}{r!(m-r)!} \quad .$$

The TI-83/84 Plus will compute combinations for us. It uses the symbol $_n C_r$ to represent a combination.

Example 4.34   In order to recruit new members, a compact disc club advertises a special introductory offer: A new member agrees to buy one compact disc at regular club prices and receives free any four compact discs of his or her choice from a collection of 69 compact discs. How many possibilities does a new member have for the selection of the four free compact discs?

Solution: We need to find $_{69}C_4$ on the TI-83/84 Plus.

1.   On the home screen, enter the total number of objects, here 69.

2. Press MATH and arrow over to the PRB menu. See Figure 4.8.

3. Press 3 or arrow down to **3:nCr** and press ENTER. The calculator will return to the home screen.

4. Enter the number of objects selected, here 4 and press ENTER. Your calculator will compute the answer and your screen will appear as in Figure 4.9.

Figure 4.8

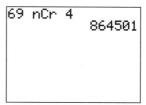
Figure 4.9

4.8 Practice Problems

Problem 4.182   Investment firms usually have a large selection of mutual funds which an investor can choose. One such firm has 30 mutual funds. Suppose that you plan to invest in four of these mutual funds, one during each quarter of the next year. In how many different ways can you make the four investments?

Problem 4.185   At a movie festival, a team of judges is to pick the first, second and third place winners from the 18 films entered. How many possibilities are there?

Problem 4.186 The sales manager of a clothing company needs to assign seven salespeople to seven different territories. How many possibilities are there for the assignments?

CHAPTER 5
DISCRETE RANDOM VARIABLES*

LESSON 5.1   DISCRETE RANDOM VARIABLES AND PROBABILITY DISTRIBUTIONS*

In this section we introduce discrete random variables and probability distributions.

Definition 5.1   A **random variable** is a quantitative variable whose value depends on chance.

Definition 5.2   A **discrete random variable** is a random variable whose possible values form a finite (or countably infinite) set of numbers.

Definition 5.3   **Probability distribution:** A listing of the possible values and corresponding probabilities of a discrete random variable; or a formula for the probabilities.

> **Probability histogram:** A graph of the probability distribution that displays the possible values of a discrete random variable on the horizontal axis and the probabilities of those values on the vertical axis. The probability of each value is represented by a vertical bar whose height is equal to the probability.

Example 5.2   Professor Weiss asked his introductory statistics students to state how many siblings they have. Table 5.1 presents a grouped-data table for that information. The table shows, for instance, that 11 of the 40 students have two siblings. Let X denote the number of siblings of a randomly selected student.

    a. Determine the probability distribution of the random variable X.

    b. Construct a probability histogram for the random variable X.

Table 5.1	Siblings X	Frequency f
	0	8
	1	17
	2	11
	3	3
	4	1
		40

Solution: To determine the probability of each of the possible values of the random variable X, we use the f/N rule. We need to divide each of the frequencies by 40.

1. With the possible values of X in L_1 and the corresponding frequencies in L_2, from the home screen press 2nd L2 ÷ 40 STO▶ 2nd L3 ENTER. Your screen should appear as in Figure 5.1.

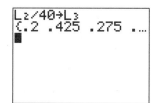

Figure 5.1

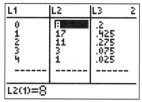

Figure 5.2

Your probabilities are now stored in L_3. You may view the values by scrolling using the arrow keys on the home screen, or by entering the list editor. See Figure 5.2.

Now we will construct the histogram. Be sure all other graphs are off including ⌈Y=⌉ plots.

1. Press ⌈2nd⌉ STAT PLOT ⌈ENTER⌉ to access Plot 1.

2. Turn Plot 1 on by highlighting ON and pressing ⌈ENTER⌉.

3. Select the histogram by highlighting the ⊞ icon and pressing ⌈ENTER⌉.

4. For Xlist, we wish to use our random variable, so for this example use L₁ (press ⌈2nd⌉ L1 ⌈ENTER⌉).

5. For frequency, we wish to use our probabilities, so for this example use L₃ (press ⌈2nd⌉ L3 ⌈ENTER⌉). Your screen should appear as in Figure 5.3.

6. Now we must set our window. Recall that ZoomStat does not work for histograms. Press ⌈WINDOW⌉ to access the window menu. Set Xmin at 0.5 below your lowest x value and Xmax at 0.5 above your highest x value. Set Xscl as 1. Ymin as -0.01, Ymax as 0.5, and Yscl as 0.05. Note that Ymax may need to be set higher if you have a probability greater than 0.5. See Figure 5.4 for the window settings for our example.

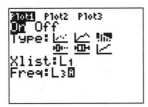
Figure 5.3

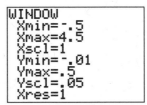
Figure 5.4

7. Press ⌈GRAPH⌉ to see the graph of your histogram. See Figure 5.5. If you do not wish to view the axis, press ⌈2nd⌉ FORMAT and arrow to AxesOff and press ⌈ENTER⌉. Press ⌈GRAPH⌉. See Figure 5.6.

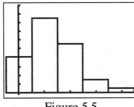
Figure 5.5

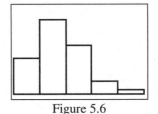
Figure 5.6

You can view the classes and the probabilities by pressing ⌈TRACE⌉ and using the arrow keys to scroll across the screen.

5.1 Practice Problems

Problem 5.7 The National Aeronautics and Space Administration (NASA) compiles data on space shuttle launches and publishes them on their Web site. Table 5.2 displays a frequency distribution for the number of crew members on each shuttle mission from April, 1981, to July, 2000.

    a. Determine the probability distribution of the random variable crew size.

    b. Construct a probability histogram for the random variable crew size.

Table 5.2	Crew Size	2	3	4	5	6	7	8
	Frequency	4	1	2	36	18	33	2

Problem 5.8   From the document *American Housing Survey for the United States*, published by the U.S. Census
Bureau, we obtained the following frequency distribution for the number of people per occupied
housing unit, where we have used "7" in place of "7 or more" Frequencies are in millions of housing
units. Determine the probability distribution.

Table 5.3	No. of people	1	2	3	4	5	6	7
	Frequency	27.9	34.4	17.0	15.5	6.8	2.3	1.4

Problem 5.9   The *Television Bureau of Adverting, Inc* publishes information on color television ownership in
Trends in Television. Table 5.4 contains the probability distribution for the number of color
televisions owned by a randomly selected household with annual income between $15,000 and
$29,999.

Table 5.4	No. of TV's	0	1	2	3	4	5
	Probability	0.009	0.376	0.371	0.167	0.061	0.016

a. Determine the probability the number of color TVs owned is 1 or more.

b. Determine the probability the number of color TVs owned is 2.

c. Determine the probability the number of color TVs owned is between 1 and 3, inclusive.

d. Determine the probability the number of color TVs owned is 1 or 3 or 5.

LESSON 5.2   THE MEAN AND STANDARD DEVIATION OF A DISCRETE RANDOM VARIABLE*

Definition 5.4   The **mean of a discrete random variable X** is denoted by μ_X or, when no confusion will arise,
simply by μ. It is defined by

$$\mu = \sum \left[xP(X = x) \right].$$

The terms **expected value** and **expectation** are commonly used in place of *mean*.

Definition 5.5   The **standard deviation of a discrete random variable X** is denoted by σ_X, or when no confusion
will arise, simply by σ. It is defined by

$$\sigma = \sqrt{\sum \left[(x - \mu)^2 P(X = x) \right]}.$$

The standard deviation of a discrete random variable can also be obtained from the computing formula

$$\sigma = \sqrt{\sum \left[x^2 P(X = x) - \mu^2 \right]}.$$

The square of the standard deviation, σ^2, is called the **variance** of the random variable X.

The TI-83/84 Plus will compute the mean and standard deviation of a discrete random variable.

Examples 5.7/5.8 Prescott National Bank has six tellers available to seven customers. The number of tellers busy with customers at, say, 1:00 p.m. varies from day to day and depends on chance; so it is a random variable, which we call X. Past records indicate that the probability distribution X, is as shown in Table 5.5.

a. Find the mean of the random variable X.

b. Find the standard deviation of the random variable X.

Table 5.5

X	P(X = x)
0	0.029
1	0.049
2	0.078
3	0.155
4	0.212
5	0.262
6	0.215
	1.000

Solution:

1. With the x values in L_1 and the corresponding probabilities in L_2, press $\boxed{\text{STAT}}$ and arrow over to the CALC menu.

2. Press $\boxed{1}$ or press $\boxed{\text{ENTER}}$ for **1-Var Stats**. This will now appear on your home screen.

3. Enter the x list first and then the frequency list. For this example, we will use L_1 and L_2. Be sure to separate the names of the lists with a comma (key above the $\boxed{7}$ key). Your screen should appear as in Figure 5.7.

4. Press $\boxed{\text{ENTER}}$. The results will be displayed as in Figure 5.8. Note that the TI-83/84 Plus only uses one symbol for the mean so our μ is listed as x-bar.

Figure 5.7

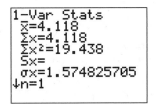

Figure 5.8

The mean number of tellers busy with customers at 1:00 p.m. is 4.118. Of course, there will never be a day when 4.118 tellers are busy with customers at 1:00 p.m. The mean of 4.118 simply indicates that over the course of many days, the average number of busy tellers at 1:00 p.m. will be about 4.118. The standard deviation of the number of tellers busy with customers at 1:00 p.m. is 1.6.

5.2 Practice Problems

Problem 5.21 The National Aeronautics and Space Administration (NASA) compiles data on space shuttle launches and publishes them on their Web site. Table 5.6 displays the probability distribution for the number of crew members on each shuttle mission from April, 1981, to July, 2000.

Table 5.6	Crew Size	2	3	4	5	6	7	8
	Probability	0.042	0.010	0.021	0.375	0.188	0.344	0.021

a. Find the mean of the random variable crew size.

b. Find the standard deviation of the random variable crew size.

Problem 5.22 From the document *American Housing Survey for the United States*, published by the U.S. Census Bureau, we obtained the following frequency distribution for the number of people per occupied housing unit, where we have used "7" in place of "7 or more." Frequencies are in millions of housing units. Determine the probability distribution.

Table 5.7	No. of people	1	2	3	4	5	6	7
	Frequency	0.265	0.327	0.161	0.147	0.065	0.022	0.013

a. Find the mean of the random variable number of people per housing unit.

b. Find the standard deviation of the random variable number of people per housing unit.

Problem 5.23 The *Television Bureau of Adverting, Inc* publishes information on color television ownership in *Trends in Television*.   Table 5.8 contains the probability distribution for the number of color televisions owned by a randomly selected household with annual income between $15,000 and $29,999.

Table 5.8	No. of TV's	0	1	2	3	4	5
	Probability	0.009	0.376	0.371	0.167	0.061	0.016

a. Find the mean of the random variable TV's.

b. Find the standard deviation of the random variable TV's.

LESSON 5.3   THE BINOMIAL DISTRIBUTION*

Formula 5.1: Let X denote the total number of successes in n Bernoulli trials with success probability p. Then the probability distribution of the random variable X is given by the formula

$$P(X = x) = \binom{n}{x} p^x (1 - p)^{n-x}.$$

The random variable X is called a **binomial random variable** and is said to have the **binomial distribution** with parameters n and p.

Example 5.12   According to tables provided by the U.S. National Center for Health Statistics in *Vital Statistics of the United States*, there is about an 80% chance that a person age 20 will be alive at age 65. Suppose three people age 20 are selected at random. Find the probability that the number alive at age 65 will be

a. exactly two.   b. at most one.   c. at least one.

d. Determine the probability distribution of the number alive at age 65.

Solution:

a. The TI-83/84 Plus has the binomial distribution built into it. It can find the probability of an individual x value or the entire distribution.

1. From the home screen, press [2nd] DISTR to access the distribution menu. Your screen should appear as in Figure 5.9.

2. Press the zero key or arrow down to **0:binompdf(** and press [ENTER]. Your calculator will return to the home screen and display **binompdf(**.

3. The format of the binompdf command is **binompdf(**numtrials, p, x) with the x being optional. To find our first value of exactly two, we will set up the command as **binompdf(**3, 0.8, 2) and then press [ENTER]. Your screen will appear as in Figure 5.10.

Figure 5.9

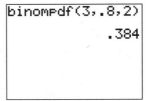

Figure 5.10

The probability that exactly two of the three people will be alive at age 65 is 0.384.

b. To determine the probability of at most one being alive at age 65, we need to know the probabilities that zero and one are alive at age 65 and sum the two probabilities. There are two ways of doing this on the TI-83/84 Plus. The first involves computing each individual probability on the calculator and then summing them.

1. From the home screen, press [2nd] DISTR to access the distribution menu. Your screen should appear as in Figure 5.9.

2. Press the zero key or arrow down to **0:binompdf(** and press [ENTER]. Your calculator will return to the home screen and display **binompdf(**.

3. The format of the binompdf command is **binompdf(** numtrials, p, x) with the x being optional. However, we can make x a list containing all the values we are interested in. To do this, use brackets to denote the list and separate each element by a comma. Note that the brackets are located above the parentheses and are accessed using the [2nd] key. It is best to store the results in a list for easy viewing and summing. Therefore, we will store the results in L_1. For this example our command would be **binompdf(**3, 0.8, {0,1}) [STO▶] [2nd] L1. Press [ENTER] and your screen should appear as in Figure 5.11.

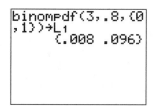

Figure 5.11

Figure 5.12

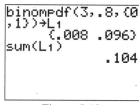

Figure 5.13

4. To sum this list and get the total probability, press [2nd] LIST and arrow over to the MATH menu. Your screen should appear as in Figure 5.12.

5. Press [5] or arrow down to **5:sum(** and press [ENTER]. Your calculator will return to the home screen and display **sum(**.

6. Enter the name of the list where you stored the probabilities. For this example we used L_1 so press [2nd] L1 [ENTER] and your screen will appear as in Figure 5.13.

The probability that exactly zero people are alive at age 65 is 0.008 and that one person is alive at age 65 is 0.096. Summing these, we get the probability 0.104 that at most one person will be alive at age 65.

The other option for this example is to compute and sum the probabilities in one step. This is appropriate for use when problems call for probabilities involving the phrase at most.

1. From the home screen, press [2nd] DISTR to access the distribution menu. Your screen should appear as in Figure 5.9.

2. Press [ALPHA] A or arrow down to **A:binomcdf(** and press [ENTER]. Your calculator will return to the home screen and display **binomcdf(**.

3. The format of the binomcdf command is **binomcdf(** numtrials, p, x). For this example, our command will be **binomcdf(** 3, 0.8, 1) [ENTER]. Your screen will appear as in Figure 5.14.

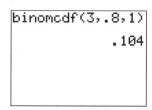

Figure 5.14

The probability that at most one person will be alive at age 65 is 0.104.

 c. To determine the probability of at least one person being alive at age 65, we need to know the probabilities that one, two and three are alive at age 65 and sum the three probabilities. There are two ways of doing this on the TI-83/84 Plus. The first method is to follow the first set of instructions in the last example and compute each individual probability on the calculator and then sum them. Although this is not hard on our example, it can be tedious when n is large.

The other method can be used for any at least "x" problem. Recall that at least x is the complement of at most x - 1. Therefore, we can use binomcdf and the concept of the complement. Our command becomes 1 - **binomcdf(** numtrials, p, x - 1).

1. From the home screen, press ☐ ☐ 2nd DISTR to access the distribution menu. Your screen should appear as in Figure 5.9.

2. Press ALPHA A or arrow down to **A:binomcdf(** and press ENTER. Your calculator will return to the home screen and display **binomcdf(**.

3. The format of the binomcdf command is **binomcdf(** numtrials, p, x). For this example, our command will be **binomcdf(** 3, 0.8, 1 - 1) ENTER. Your screen will appear as in Figure 5.15.

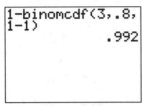

Figure 5.15

   d. To determine the entire probability distribution, leave off the x portion of the command. The calculator will automatically compute the probability for all possible x values. Because this can sometimes be hard to read by scrolling on the home screen, it is recommended that you store the probabilities in a list for easy viewing.

1. From the home screen, press 2nd DISTR to access the distribution menu. Your screen should appear as in Figure 5.9.

2. Press the zero key or arrow down to **0:binompdf(** and press ENTER. Your calculator will return to the home screen and display **binompdf(**.

3. The format of the binompdf command is **binompdf(**numtrials, p) and we need to store this into a list. For this example we will use L_1. Therefore our command is

binompdf(3, 0.8) 2nd STO▸ L1. Press ENTER and your screen should appear as in Figure 5.16. By entering the statistics list editor you can view all the values.

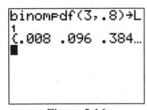

Figure 5.16

5.3 Practice Problems

Problem 5.48   The National Institute of Mental Health reports that there is a 20% chance of an adult American suffering from a psychiatric disorder. Four randomly selected adult Americans are examined for a psychiatric disorder. If a success corresponds to an adult American having a psychiatric disorder, find the probability that the number with a psychiatric disorder is

   a. exactly one.   b. at most three.   c. at least one.

Problem 5.63   Traffic Fatalities and Intoxication. The National Safety Council publishes information about automobile accidents in *Accident Facts*. According to that document, the probability is 0.40 that a traffic fatality will involve an intoxicated or alcohol impaired driver or nonoccupant. In eight traffic fatalities, find the probability that the number, Y, that involve an intoxicated or alcohol-impaired driver or nonoccupant is

    a. exactly three; at least three; at most three.
    b. between two and four, inclusive.

Problem 5.67   According to the Centers for Disease Control and Prevention publication *Health, United States*, in 2002, 16.5% of persons under the age of 65 had no health insurance coverage. Suppose that, today, four persons under the age of 65 are randomly selected.

    a. Assuming that the uninsured rate is the same today as it was in 2002, determine the probability distribution for the number, X, who have no health insurance coverage.
    b. Determine and interpret the mean of X.

Simulating a Binomial Random Variable
Sometimes it is desirable to simulate a random variable rather than taking an actual sample. The TI-83/84 Plus has a random number generator specifically for binomial random variables.

Example 5.4   Simulate 500 observations of the number of heads obtained in three tosses of a balanced dime. [†]

Solution: This is a binomial distribution where n = 3 and the probability of a success is the probability of getting a head so p = 0.5.

    1. From the home screen, press MATH and arrow to the PRB menu. See Figure 5.17.

    2. Press 7 or arrow down to **7:randBin(** and press ENTER. Your calculator will return to the home screen with **randBin(** displayed.

    3. The format of this command is **randBin**(numtrials, prob,[,numsimulations]) with the number of simulations being optional. If the number of simulations is not specified the calculator assumes it to be one. For this example, our number of trials is 3, with a probability of success of 0.5, and we want to do 500 simulations. Thus our command for this example is **randBin**(3, 0.5, 500). It is a good idea to store the simulations in a list for easy reference, so we will store them in L_1 by pressing STO▶ 2nd L1. Then press ENTER. See Figure 5.18.

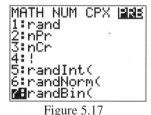
Figure 5.17

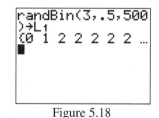
Figure 5.18

The numbers of heads obtained in three tosses of a balanced dime for 500 observations are now stored in L_1. Now we must tally them. We will do this by summing the number of each value in the list using the calculator.

[†] We have chosen to use only 500 observations rather than the 1000 that *Introductory Statistics* and *Elementary Statistics* use because of the limits on memory that the TI-83/84 Plus has. Also, this simulation will take about 1 minute to run on the TI-83/84 Plus.

4. From the home screen, press 2nd LIST, and arrow to the MATH menu. See Figure 5.19.

5. Press ∑ 5 or arrow down to **5:sum(** and press ENTER.

6. Press 2nd L1 to indicate L_1, 2nd TEST ENTER to get the =, 0 to sum the zeros, and) STO▸ 2nd L2 ((1)) to store the result in the first element of L_2. See Figure 5.20.

Figure 5.19

Figure 5.20

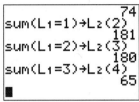
Figure 5.21

7. Press ENTER. The total number of zeros will be displayed. Note that your results will almost certainly differ from the ones shown.

8. Press 2nd ENTRY to display your last command and then use your arrow keys to move around the command and change it to sum the ones and store them in the second element of L_2. Press ENTER.

9. Repeat until you have summed all the values up to and including the number of trials. Here that is three. The results are shown in Figure 5.21.

10. To get the percentages we need to divide the frequencies by 500. Press 2nd L2 ÷ 500 STO▸ 2nd L3 ENTER to store the percentages in L_3. See Figure 5.22.

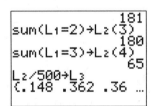
Figure 5.22

Although your values will almost certainly be different, the end result will be similar. Compare your results to the actual probabilities listed in Table 5.9.

Table 5.9	Number of heads, x	P(X = x)
	0	0.125
	1	0.375
	2	0.375
	3	0.125
		1.000

LESSON 5.4   THE POISSON DISTRIBUTION*

Formula 5.3: Probabilities for a random variable X having a Poisson distribution are given by the formula

$$P(X = x) = e^{-\lambda} \frac{\lambda^x}{x!}, \quad x = 0, 1, 2, \dots,$$

where λ is a positive real number and $e \approx 2.718$. The random variable X is called a **Poisson Random Variable** and is said to have the **Poisson Distribution** with parameter λ.

Example 5.16   Desert Samaritan Hospital, located in Mesa, Arizona, keeps records of emergency-room traffic. From these records we find that the number of patients arriving between 6:00 p.m. and 7:00 p.m. has a Poisson distribution with parameter $\lambda = 6.9$. Determine the probability that, on a given day, the number of patients arriving at the emergency room between 6:00 p.m. and 7:00 p.m. will be

a. exactly four.

b. at most two.

c. between four and 10, inclusive.

Solution:

a. The TI-83/84 Plus has built in functions for computing Poisson probabilities.

1. From the home screen, press [2nd] DISTR to access the distribution menu. Your screen should appear as in Figure 5.23.

2. Press [ALPHA] B or arrow down to **B:poissonpdf(** and press [ENTER]. Your calculator will return to the home screen and display **poissonpdf(**.

3. The format for this command is **poissonpdf(μ, x)**. For the Poisson distribution, μ = λ. For this example, our command will be **poissonpdf(6.9, 4)** [ENTER]. See Figure 5.24.

Figure 5.23

Figure 5.24

The probability that exactly four patients will arrive at the emergency room between 6:00 p.m. and 7:00 p.m. is 0.0952 rounded to four decimal places.

b. For this part, we want the probability of at most two arrivals that is two or fewer. We will use the poissoncdf for probabilities involving the phrase at most.

1. From the home screen, press [2nd] DISTR to access the distribution menu. Your screen should appear as in Figure 5.23.

2. Press ALPHA C or arrow down to **C:poissoncdf(** and press ENTER. Your calculator will return to the home screen and display **poissoncdf(.**

3. The format for this command is **poissoncdf(** μ, x). For the Poisson distribution, μ = λ. For this example, our command will be **poissoncdf(** 6.9, 2) ENTER. See Figure 5.25.

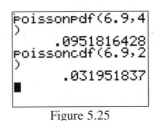

Figure 5.25

The probability of at most two arrivals between 6:00 p.m. and 7:00 p.m. is 0.032 rounded to three decimal places.

> c. We need to determine the probability of between four and 10 arrivals, inclusive. There are several options for obtaining that probability. One way is to obtain the individual probabilities and then sum those probabilities.

1. From the home screen, press 2nd DISTR to access the distribution menu. Your screen should appear as in Figure 5.23.

2. Press ALPHA B or arrow down to **B:poissonpdf(** and press ENTER. Your calculator will return to the home screen and display **poissonpdf(.**

3. The format for this command is **poissonpdf(** μ, x). For the Poisson distribution, μ = λ. We can use a list of values instead of a single x value. To do this, use brackets to denote the list and separate each element by a comma. Note that the brackets are located above the parentheses and are accessed using the 2nd key. It is best to store the results into a list for easy viewing and summing. Therefore we will store the results in L_1. For this example our command will be **poissonpdf(** 6.9, {4, 5, 6, 7, 8, 9, 10}) STO▸ 2nd L1 ENTER. See Figure 5.26.

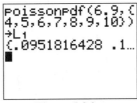   

Figure 5.26   Figure 5.27

4. To sum this list to get the total probability, press 2nd LIST and arrow over to the MATH menu. Your screen should appear as in Figure 5.27.

5. Press 5 or arrow down to **5:sum(** and press ENTER.   Your calculator will return to the home screen and display **sum(.**

6. Enter the name of the list where you stored the probabilities. For this example we used L_1 so press 2nd L1 ENTER and your screen will appear as in Figure 5.28.

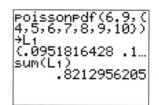

Figure 5.28

The probability of between four and 10 arrivals is 0.8213 rounded to four decimal places. There are other more efficient ways to do this calculation. Use the poissoncdf function instead of the poissonpdf. This exercise is left to the reader.

5.4 Practice Problems

Problem 5.84   Based on past records, the owner of a fast-food restaurant knows that, on the average, 2.4 cars use the drive-through window between 3:00 P.M. and 3:15 P.M. Assuming the number of such cars has a Poisson distribution, obtain the probability that between 3:00 P.M. and 3:15 P.M., the number of cars using the drive-through window is

a. exactly two.   b. at least three.   c. at most four.

Problem 5.85   A 1910 article, "The Probability Variations in the Distribution of α Particles," appearing in *Philosophical Magazine*, describes the results of experiments with polonium. The experiments, conducted by Ernest Rutherford and Hans Geiger, indicate that the number of α particles reaching a small screen during an 8-minute interval has a Poisson distribution with parameter $\lambda = 3.87$. Determine the probability that, during an 8-minute interval, the number of α particles reaching the screen will be

a. exactly four.   b. at most one.   c. between two and five, inclusive.

Problem 5.87   A paper by L.F. Richardson, published in the *Journal of Royal Statistical Society*, analyzed the distribution of wars in time. From the data, we find that the number of wars that begin during a given calendar year has approximately a Poisson distribution with parameter $\lambda = 0.7$. If a calendar year is selected at random, find the probability that the number of wars that begin during that calendar year will be

a. zero.   b. at most two.   c. between one and three, inclusive.

CHAPTER 6
THE NORMAL DISTRIBUTION

LESSON 6.1 INTRODUCING NORMALLY DISTRIBUTED VARIABLES

A variable is said to be **normally distributed** or to have a **normal distribution** if its distribution has the shape of a normal curve.

A normally distributed variable having mean 0 and standard deviation 1 is said to have the **standard normal distribution**. Its associated normal curve is called the **standard normal curve**.

The standardized version of a normally distributed variable x,

$$z = \frac{x - \mu}{\sigma},$$

has the standard normal distribution.

Example 6.2   Gestation periods of humans are normally distributed with a mean of 266 days and a standard deviation of 16 days. Use the TI-83/84 Plus to simulate 500 human gestation periods and then obtain a histogram of the results. (Note: the TI-83/84 Plus will handle up to 999 observations but memory problems may result.)

Solution: The gestation period is the variable x and, for humans, it is normally distributed with a mean of 266 days and a standard deviation of 16 days. We will use the TI-83/84 Plus to simulate 500 observations of the gestation period of humans.

1. Press MATH

2. Arrow over to PRB.

3. Select **randNorm(** by pressing ⑥ or arrow down to 6 and pressing ENTER. See Figure 6.1. The command will appear on your home screen.

4. The format for this command is **randNorm(** μ, σ , number of trials) where number of trials is the number of observations you want. If not specified, the default number of trials is 1.

5. Store the generated data in List 1 by pressing STO▸ 2nd L1 ENTER.

 Note: The TI-83/84 Plus will need some time to do the simulation. The calculator indicates it is still working by a small vertical moving line in the upper right of the screen.

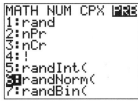
Figure 6.1

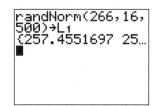
Figure 6.2

As a result, we now have 500 simulated observations of human gestation periods stored in List 1.

5. Now construct a histogram. Rather than setting the window ourselves we will use ZoomStat (ZOOM 9) to let the calculator pick the window. See Figure 6.3.

Your simulation will give slightly different results, but the results should be similar.

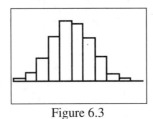

Figure 6.3

6.1 Practice Problems

Problem 6.23   According to the National Health and Nutrition Examination Survey, the serum (noncellular portion of blood) total cholesterol level of U.S. females 20 years old or older is normally distributed with a mean of 206 mg/dL (milligrams per deciliter) and a standard deviation of 44.7 mg/dL. Simulate 500 observations of serum total cholesterol level of U.S. females 20 years old or older and then obtain a histogram of the results.

Problem 6.31   One of the larger species of tarantulas is the *Grammostola mollicoma*, whose common name is the Brazilian Giant Tawny Red. A tarantula has two body parts. The anterior part of the body is covered above by a shell, or carapace. From a recent article, by F. Costa and F. Perez-Miles titled, "Reproductive Biology of Uruguayan Theraphosids" (*The Journal of Arachnology*, Vol. 30, No. 3, pp. 571-587) we find that the carapace length of the adult male of *G. mollicoma* is normally distributed with a mean of 18.14 mm and a standard deviation of 1.76 mm. Simulate 500 carapace lengths of adult male *G. mollicoma* and then obtain a histogram of the results.

Problem 6.33   As reported in *Runner's World* magazine, the times of the finishers in the New York City 10 km run are normally distributed with a mean of 61 minutes and a standard deviation of 9 minutes. Simulate the times of 500 finishers of the New York City 10 km run and then obtain a histogram of the results.

LESSON 6.2   AREAS UNDER THE STANDARD NORMAL CURVE

Recall that the standard normal curve is the graph of the normal distribution with mean 0 and standard deviation 1. It is also sometimes called the z-curve. The following properties hold:

Property 1: The total area under the standard normal curve is equal to 1.

Property 2: The standard normal curve extends indefinitely in both directions, approaching, but never touching, the horizontal axis as it does so.

Property 3: The standard normal curve is symmetric about 0.

Property 4: Almost all the area under the standard normal curve lies between -3 and 3.

Example 6.3   Determine the area under the standard normal curve that lies to the left of 1.23.

Solution: We need to find the area under the curve from $-\infty$ to 1.23. To do this we will use the **normalcdf(** function on the TI-83/84 Plus.

1. Press 2nd DISTR to access the distribution menu. See Figure 6.4.

2. Select **normalcdf(** by pressing 2 or arrow down to **2:normalcdf(** and pressing ENTER.

3. The format of this command is **normalcdf**(lowerbound, upperbound, μ, σ) with the default mean of 0 and the default standard deviation of 1. Because the TI-83/84 Plus does not understand $-\infty$ we will use -1 times 10 raised to the 99th power or -1E99. Thus our command becomes **normalcdf**(-1E99,1.23) ENTER . Note: The E is obtained by pressing 2nd EE. See Figure 6.5 for the command and the answer.

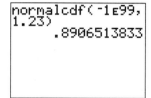

Figure 6.4   Figure 6.5

We can also use this command to find the area to the right of a z-score and the area between two z-scores. The following examples illustrate this.

Example 6.4   Determine the area under the standard normal curve that lies to the right of 0.76.

Solution: This time we need to find the area under the curve from 0.76 to ∞.

    1. Access the **normalcdf**(command by pressing 2nd DISTR 2 .

    2. We will use 1E99 to estimate ∞. Thus our command becomes **normalcdf**(0.76, 1E99). The command and answer are shown in Figure 6.6.

Example 6.5   Determine the area under the standard normal curve that lies between -0.68 and 1.82.

Solution:

    1. Access the **normalcdf**(command by pressing 2nd DISTR 2 .

    2. Our command is **normalcdf**(-0.68,1.82). The command and answer are shown in Figure 6.6.

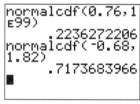

Figure 6.6

Note that the TI-83/84 Plus gives values to the tenth decimal place unless you have set your mode for something else. The tables in the textbook only use four places. Therefore, your answers may differ slightly from the textbook unless you round to four decimal places.

6.2 Practice Problems

Problem 6.55   Determine the area under the standard normal curve that lies to the left of
      a) 2.24   b) -1.56   c) 0   d) -4

Problem 6.57   Determine the area under the standard normal curve that lies to the right of
      a) -1.07   b) 0.6   c) 0   d) 4.2

Problem 6.59   Determine the area under the standard normal curve that lies between
   a) -2.18 and 1.44   b) -2 and -1.5
   c) 0.59 and 1.51   d) 1.1 and 4.2

Finding a Z-score for a Specified Area
Up until now, we have used the TI-83/84 Plus to find the areas under the standard normal curve. Now we will use the TI-83/84 Plus to find the z-scores corresponding to a specified area under the standard normal curve.

Example 6.6   Determine the z-score having area 0.04 to its left under the standard normal curve.

Solution: We will use the **invNorm(** function to find the z-score.

   1. Press 2nd DISTR to access the distribution menu. See Figure 6.7.

   2. Select **invNorm(** by pressing 3 or arrow down to **3:invNorm(** and pressing ENTER.

   3. The format of this command is **invNorm(**area to the left, μ, σ) Like the normalcdf(command the default mean and standard deviation are 0 and 1 respectively. Thus, our command is **invNorm(**0.04)
   ENTER. The command and its output are shown in Figure 6.8.

Figure 6.7

Figure 6.8

The TI-83/84 Plus computes the z-score to 9 decimal places. Therefore your answers may differ slightly from the textbook even when rounding the final answer to two decimal places.

z_α **Notation**

z_α is used to denote the z-score having area alpha, α to its right under the standard normal curve.

Example 6.7   Use the TI-83/84 Plus to find
   a. $z_{0.025}$ b. $z_{0.05}$

Solution: We will use the **invNorm(** command. However, it only accepts area to the left and we have been given the area to the right. Therefore, we must start by finding the area to the left. Recall property 1, the area under the standard normal curve is equal to 1. To find the area to the left, subtract the area to the right from 1. For part a, this means 1- 0.025 = 0.975 and for part b, this means 1- 0.05 = 0.95. Our commands become **invNorm(**0.975) and **invNorm(**0.95), respectively. See Figure 6.9 for the commands and answers.

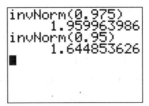
Figure 6.9

Note that for part a, when rounding to two decimal places we get the same answer as we do using the table in the textbook. However, for part b, we do not get the same answer unless we round to three decimal places. This rounding can change the final answers for some problems later in the book.

Example 6.8   Determine the two z-scores that divide the area under the standard normal curve into a middle 0.95 area and two outside 0.025 areas.

Solution: If the outside areas are 0.025, that means that the area to the left of one z-score is 0.025 and the area to the right of the other z-score is 0.025. Therefore, to find the first z-score use the command **invNorm**(0.025) and for the second use the command **invNorm**(0.975) where 0.975 is 1 minus the area to the right. The results are shown in Figure 6.10.

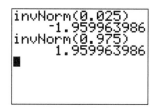

Figure 6.10

Note: we can also solve the above problem by first finding the z-score on the left, which is -1.96. Using the symmetry property, obtain the z-score on the right, 1.96.

6.2 Practice Problems (continued)

Problem 6.67   For the standard normal curve, determine the z-score for which the area to its left is 0.025.

Problem 6.69   For the standard normal curve, determine the z-score with an area of 0.75 to its left.

Problem 6.71   For the standard normal curve, determine the z-score having area 0.95 to its right.

Problem 6.73   Find $z_{0.33}$

Problem 6.75   Find
           a. $z_{0.03}$ b. $z_{0.005}$

Problem 6.77   Determine the two z-scores that divide the area under the curve into a middle 0.90 area and two outside 0.05 areas.

LESSON 6.3   WORKING WITH NORMALLY DISTRIBUTED VARIABLES

To determine a percentage or probability for a normally distributed variable, we must first sketch the normal curve associated with the variable and determine the region of interest. Then we can use the **normalcdf(** command to change our x's into z-scores and compute the area.

Example 6.9   Intelligence quotients (IQs) measured on the Stanford Revision of the Binet-Simon Intelligence Scale are known to be normally distributed with a mean of 100 and a standard deviation of 16. Obtain the percentage of people having IQs between 115 and 140.

Solution: Recall **normalcdf**(lowerbound, upperbound, μ, σ) is the format of the command. For our problem, the lowerbound is 115, the upperbound is 140 and the mean and standard deviation are 100 and 16, respectively.

1. Access the **normalcdf(** command by pressing [2nd] DISTR [2].

2. Your command should be **normalcdf**(115,140,100,16) [ENTER]. The command and its result are shown in Figure 6.11.

Therefore we can say that 16.80% of all people have IQs between 115 and 140. We can interpret this result probabilistically by saying that the probability is 0.1680 that a randomly selected person will have an IQ between 115 and 140.

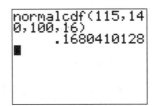

Figure 6.11

The textbook gives the answer to this example as 0.1674. This is because the textbook must round the z-scores to two decimal places to match the tables and the areas are rounded to four decimal places in the tables. The TI-83/84 Plus does not round the z-scores or the areas. Therefore, answers will almost always be slightly different when using this method. To obtain answers closer to the text book, compute the z-scores first, then use the methods of Lesson 6.2 to obtain the areas using the z-scores.

Recall that we must use -1E99 to represent $-\infty$ and 1E99 to represent ∞.

Finding Observations for a Specified Percentage
We can use the following method to obtain quartiles, deciles or any other percentile for a normally distributed variable.

Example 6.11   Obtain the 90th percentile for IQs.

Solution: The 90th percentile is the IQ that is higher than those of 90% of all people. This means that our desired x-value has an area 0.90 to its left under the normal curve associated with IQs. To find this value, use **invNorm(**.

1. Press [2nd] DISTR [3] to access the command.

2. Remember the format is **invNorm**(area to the left, μ, σ). Our command is **invNorm**(0.90,100,16). The command and its result are shown in Figure 6.12.

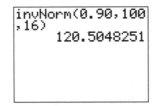

Figure 6.12

Once again, the calculator gives a slightly different answer than the table because the table must round the z-score.

6.3 Practice Problems

Problem 6.93   One of the larger species of tarantulas is the *Grammostola mollicoma*, whose common name is the Brazilian Giant Tawny Red. A tarantula has two body parts. The anterior part of the body is covered above by a shell, or carapace. From a recent article, by F. Costa and F. Perez-Miles titled, "Reproductive Biology of Uruguayan Theraphosids" (*The Journal of Arachnology*, Vol. 30, No. 3, pp. 571-587) we find that the carapace length of the adult male of *G. mollicoma* is normally distributed with a mean of 18.14 mm and a standard deviation of 1.76 mm.

   a. Find the percentage of adult male *G. mollicoma* that have carapace length between 16 mm and 17mm.
   b. Find the percentage of adult male *G. mollicoma* that have carapace length exceeding 19 mm.
   c. Determine and interpret the quartiles for carapace length of the adult male *G. mollicoma*.
   d. Obtain and interpret the 95th percentile for carapace length of the adult male *G. mollicoma*.

Problem 6.95   As reported by *Runner's World* magazine, the times of the finishers in the New York City 10-km run are normally distributed with a mean of 61 minutes and a standard deviation of 9 minutes. Let X be the time of a randomly selected finisher. Find

   a. Determine the percentage of finishers with times between 50 and 70 minutes.
   b. Determine the percentage of finishers with times less than 75 minutes.
   c. Obtain the 40th percentile for the finishing times.
   d. Find the 8th decile for the finishing times.

Problem 6.97   An article by Scott M. Berry entitled "Drive for Show and Putt for Dough" (*Chance*, Vol. 12(4), pp. 50-54) discussed driving distances of PGA players. The mean distance for tee shots in the 1999 men's PGA tour is 272.2 yards with a standard deviation of 8.12 yards. Assuming that the 1999 tee-shot distances are normally distributed, find the percentage of such tee shots that went

   a. between 260 and 280 yards.   b. more than 300 yards.

Problem 6.105 *Opisthotrochopodus n. sp.* is a polychaete worm that inhabits deep sea hydrothermal vents among the Mid-Atlantic Ridge. According to an article by Van Dover, et.al. in *Marine Ecology Progress Series* (Vol. 181, pp. 201-214), the lengths of female polychaete worms are normally distributed with mean of 6.1mm and standard deviation of 1.3 mm. Let *X* denote the length of a randomly selected female polychaete worm. Find

   a. $P(X \leq 3)$.   b. $P(5 < X < 7)$.

LESSON 6.4   ASSESSING NORMALITY; NORMAL PROBABILITY PLOTS

How do we decide whether or not a variable is normally distributed, or at least approximately so, based on a sample of observations? This decision is the key for subsequent analyses like percentage or percentile calculations and statistical inferences.

For large samples, a histogram of the observations should be roughly bell-shaped. For small samples, it is often hard to decide if a histogram is bell-shaped or not. This is true for stem-and-leaf diagrams as well as dotplots. Therefore, we use a normal probability plot, which is a more sensitive technique.

A **normal probability plot** is a plot of the observed values of the variable versus the normal scores– the observations we would expect to get for a variable having the standard normal distribution. If the variable is normally distributed, then the normal probability plot should fall roughly in a straight line.

The decision of what is a roughly straight line is a subjective one and we must remember that we are using a sample to make a judgment about the population.

Guidelines for Assessing Normality Using a Normal Probability Plot

1. If the plot is roughly linear, then accept as reasonable that the variable is approximately normally distributed.

2. If the plot shows systematic deviations from linearity (e.g., if it displays significant curvature), then conclude that the variable is probably not approximately normally distributed.

Example 6.14   The Internal Revenue Service publishes data on federal individual income tax returns in *Statistics of Income, Individual Income Tax Returns*. A random sample of 12 returns from last year revealed the adjusted gross incomes, in thousands of dollars, shown in Table 6.1 Construct a normal probability plot for these data and assess the normality of adjusted gross incomes.

Table 6.1			
9.7	93.1	33.0	21.2
81.4	51.1	43.5	10.6
12.8	7.8	18.1	12.7

Solution: The population consists of all last year's federal individual income tax returns and the variable we are considering is adjusted gross income. Enter the data from Table 6.1 into List 1. The TI-83/84 Plus has a built in normal probability plot function. When using this function, the order of the data does not matter.

1. Press [2nd] STAT PLOT.

2. Be sure all plots except Plot 1 are turned off and then press [ENTER].

3. Highlight On and press [ENTER].

4. For Type, choose the last one ⌐ by moving over to it and pressing [ENTER].

5. Enter the data list as List 1 by moving to Data List and pressing [2nd] L1.

6. Choose the data axis to be X by highlighting X and pressing [ENTER].

7. Choose the mark you prefer to use by highlighting it and pressing [ENTER]. See Figure 6.13.

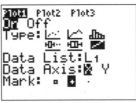

Figure 6.13

8. Set your window based on your data. (Note: you may also use the ZoomStat feature to set your window for you. Figure 6.14 is obtained using ZoomStat.) Look at your data to pick an appropriate Xmin and Xmax. For this example, set Xmin at 0, Xmax at 100 and Xscl at 10.

9. The Ymin and Ymax are a little trickier. If they are too far apart, it can change the appearance of your graph. In general, if your sample size is less than 25 you can use -2 and 2. (See Figure 6.15) If your sample size is between 25 and 50, you can use -2.5 and 2.5. Your Yscl can be set at 0.5. (See Figure 6.16 for an example of setting these values too far apart- namely at -3 and 3.)

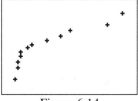
Figure 6.14

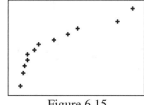
Figure 6.15

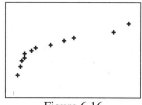
Figure 6.16

There appears to be significant curvature in Figures 6.14 and 6.15. Therefore, we conclude that adjusted gross incomes are not normally distributed.

6.4 Practice Problems

Problem 6.123 A sample of the final exam scores in a large introductory statistics course is displayed in Table 6.2. Construct a normal probability plot and assess the normality of the exam scores.

Table 6.2

88	67	64	76	86
85	82	39	75	34
90	63	89	90	84
81	96	100	70	96

Problem 6.125 Table 6.3 displays finishing times, in seconds, for the winners of fourteen 1-mile thoroughbred horse races, as found in two recent issues of *Thoroughbred Times*. Construct a normal probability plot and assess the normality of the finishing times.

Table 6.3

94.15	93.37	103.02	95.57	97.73	101.09	99.38
97.19	96.63	101.05	97.91	98.44	97.47	95.10

Problem 6.132 A paper by Shao-Chun Lu et al., entitled "LDL of Taiwanese Vegetarians are Less Oxidizable Than Those of Omnivores" (*Human Nutrition and Metabolism*, Vol. 130(6), pp. 1591-1596) compared fat consumption by vegetarians and omnivores. Table 6.4 displays the amount of fat consumed, in grams per day, for 28 vegetarians in the study. Construct a normal probability plot and assess the normality of fat consumption for Taiwanese vegetarians.

Table 6.4

20.5	31.4	35.7	52.8	27.0	40.3	45.7
19.7	32.5	33.5	58.5	30.1	61.4	33.3
35.3	54.7	54.1	56.7	35.9	58.8	25.7
66.3	35.9	35.7	47.1	38.7	16.4	42.0

Problem 6.126 The Bureau of Labor Statistics publishes information on average annual expenditures by consumers in *Consumer Expenditure Survey*. In 2005 the mean amount spent by consumers on non-alcoholic beverages was $303. A random sample of 12 consumers yielded the data in Table 6.5, in dollars, on last year's expenditures on nonalcoholic beverages. Construct a normal probability plot and assess the normality of nonalcoholic beverages expenditures.

Table 6.5

423	238	246	327	321	343
302	335	321	311	256	320

LESSON 6.5 NORMAL APPROXIMATION TO THE BINOMIAL DISTRIBUTION*

In this section, we demonstrate the approximation of binomial probabilities by using areas under a suitable normal curve.

Example 6.19 Mortality tables enable actuaries to obtain the probability that a person at any particular age will live a specified number of years. Insurance companies and others use such probabilities to determine life-insurance premiums, retirement pensions, and annuity payments. According to tables provided by the National Center for Health Statistics in *Vital Statistics of the United States*, a person of age 20 years has about an 80% chance of being alive at age 65 years. In chapter 5 we used the binomial probability formula to determine probabilities for the number of 20-year-olds out of three who will be alive at age 65. For most real-world problems, the number of people under investigation is much larger than three. Although in principle we can use the binomial probability formula to determine probabilities regardless of number, in practice we do not. Suppose, for instance, that 500 people of age 20 years are selected at random. Find the probability that

a. exactly 390 of them will be alive at age 65.
b. between 375 and 425 of them, inclusive, will be alive at age 65.

Solution :

a. The number of trials is $n = 500$, and the success probability is $p=0.80$. From the values for n and p, $np = 500 \cdot 0.8 = 400$ and $n(1 - p) = 500 \cdot 0.2 = 100$. Thus, both np and $n(1 - p)$ are greater than 5, so we can continue.

We get $\mu = 500 \cdot 0.8 = 400$ and $\sigma = \sqrt{\dfrac{0.80(1-0.80)}{500}} = 8.94$

To make the correction for continuity, we subtract 0.5 from 390 and add 0.5 to 390. Thus we need to find the area under the normal curve with parameters $\mu = 400$ and $\sigma = 8.94$ that lies between 389.5 and 390.5. Recall **normalcdf**(lowerbound, upperbound, μ, σ) is the format of the command. For our problem, the lowerbound is 389.5, the upperbound is 390.5 and the mean and standard deviation are 400 and 8.94, respectively.

1. Access the **normalcdf(** command by pressing 2nd DISTR 2.

2. Your command should be **normalcdf**(389.5,390.5,400,8.94) ENTER. The command and its result are shown in Figure 6.17.

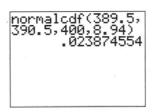

Figure 6.17

In Figure 6.17 we see the area is 0.0239 when rounded to 4 decimal places. So, *P(X = 390) = 0.0239*, approximately. The probability is about 0.0239 that exactly 390 of the 500 people selected will be alive at age 65.

b. To make the correction for continuity, we subtract 0.5 from 375 and add 0.5 to 425. Thus we need to determine the area under the normal curve with parameters $\mu = 400$ and $\sigma = 8.94$ that lies between 374.5 and 425.5. As in part (a),

1. Access the **normalcdf(** command by pressing [2nd] DISTR [2].

2. Your command should be **normalcdf(**374.5,425.5,400,8.94) [ENTER]. The command and its result are shown in Figure 6.18.

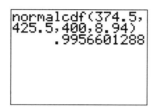

```
normalcdf(374.5,
425.5,400,8.94)
         .9956601288
```

Figure 6.18

In Figure 6.18 we see the area is 0.9957 rounded to 4 decimal places. So, $P(375 \leq X \leq 425) = 0.9957$, approximately. The probability is approximately 0.9957 that between 375 and 425 of the 500 people selected will be alive at age 65.

6.5 Practice Problems

Problem 6.145   According to the document *Current Population Survey*, published by the U.S. Census Bureau, 31.6% of U.S. adults 25 years old or older have a high school degree as their highest educational level. If 100 such adults are selected at random, determine the probability that the number who have a high school degree as their highest educational level is

a. exactly 32.
b. between 30 and 35, inclusive.
c. at least 25.

Problem 6.149   In the online *TIME* article "America's Health Checkup," A. Park reported that 40% of U.S. adults get no exercise. If 250 U.S. adults are selected at random, determine the probability that the number who get no exercise

a. is exactly 40% of those sampled.
b. exceeds 40% of those sampled.
c. is fewer than 90.

CHAPTER 7
THE SAMPLING DISTRIBUTION OF THE SAMPLE MEAN

LESSON 7.1 SAMPLING ERROR AND THE NEED FOR SAMPLING DISDTRIBUTIONS

We have discovered that using a sample to acquire information about a population is often preferable to collecting information for the entire population. However, a sample from a population provides data for only a portion of the entire population. Thus, it is unreasonable to expect the sample to yield perfectly accurate information about the population. We should anticipate that a certain amount of error will result simply because we are sampling. This error is called **sampling error**.

For example, we can choose a random sample from a population and then compare the sample mean to the population mean. The difference between the two is the sampling error associated with that sample. By repeating this process several times we can understand how the sampling error changes.

Example 7.2   Heights of Starting Players: Suppose the population of interest consists of the five starting players on the men's basketball team whom, for convenience we will call A, B, C, D, and E. Further, suppose the variable of interest is height, in inches. Table 7.1 lists the players and their heights.

Table 7.1	Player	A	B	C	D	E
	Height	76	78	79	81	86

Obtain the sampling distribution of the sample mean for samples of size 2. Find the probability that, for a random sample of size 2, the sampling error made in estimating the population mean by the sample mean will be 1 inch or less; that is, determine the probability that $\overline{x}$ will be within 1 inch of μ.

Solution: First, calculate the population mean, μ. Enter the data from Table 7.1 into List 1.

1. Press $\boxed{\text{STAT}}$.

2. Arrow over to CALC.

3. Press $\boxed{1}$ or $\boxed{\text{ENTER}}$. **1 - Var Stats** will appear on your home screen. See Figure 7.1.

4. Enter the list where the data are stored, here List 1, $\boxed{\text{2nd}}$ L1, and press $\boxed{\text{ENTER}}$. See Figure 7.2.

Figure 7.1

Figure 7.2

5. This screen yields the mean as well as several other statistics. See Figure 7.3.

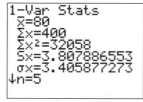

Figure 7.3

We see in Figure 7.3 that the population mean is 80.

The population contains only 5 heights so it is possible to list all the possible samples of size 2 and calculate the mean for each sample. The 10 samples of size 2 are listed in the first column of Table 7.2. The second column contains the corresponding heights and the third column contains the mean for each sample of size 2. The sampling distribution of the sample mean is the set of all possible values for the sample mean for a given sample size. Thus, the third column of Table 7.2 is the sampling distribution of the sample mean for samples of size 2.

Table 7.2

Sample	Heights	$\overline{x}$
A, B	76, 78	77.0
A, C	76, 79	77.5
A, D	76, 81	78.5
A, E	76,86	81.0
B, C	78,79	78.5
B, D	78, 81	79.5
B, E	78, 86	82.0
C, D	79, 81	80.0
C, E	79, 86	82.5
D, E	81, 86	83.5

Notice that all 10 of the sample means are relatively close to the population mean but each differs from the population mean by a small amount. This demonstrates **sampling error**. The sampling error associated with a particular sample mean can be calculated. For example, the sampling error associated with the first sample mean, 77.0 is equal to 77.0 – 80.00 = -3. Thus, a sample mean of 77.0 inches is three inches below the population mean of 80 inches.

The fourth column of Table 7.3 contains the sampling error for each possible sample mean. Negative values indicate that the sample mean was less than the population mean, whereas positive values indicate the sample mean was more than the population mean. A value of zero would indicate that the sample mean equals the population mean. Note that the sum of all the sampling errors is zero.

Table 7.3

Sample	Heights	$\overline{x}$	$\overline{x} - \mu$
A, B	76, 78	77.0	77.0 – 80.0 = -3.0
A, C	76, 79	77.5	77.5 – 80.0 = - 2.5
A, D	76, 81	78.5	78.5 – 80.0 = - 1.5
A, E	76,86	81.0	81.0 – 80.0 = 1.0
B, C	78,79	78.5	78.5 – 80.0 = - 1.5
B, D	78, 81	79.5	79.5 – 80.0 = -0.5
B, E	78, 86	82.0	82.0 – 80.0 = 2.0
C, D	79, 81	80.0	80.0 – 80.0 = 0.0
C, E	79, 86	82.5	82.5 – 80.0 = 2.5
D, E	81, 86	83.5	83.5 – 80.0 = 3.5

To find the probability that, for a random sample of size 2, the sampling error made in estimating the population mean by the sample mean will be 1 inch or less we must determine the probability that $\overline{x}$ will be within 1 inch of μ. In the third column of Table 7.3 we see that 3 of the sample means are within 1 inch of the population mean. Thus, the probability that, for a random sample of size 2, the sampling error made in estimating the population mean by the sample mean will be 1 inch or less is 3/10.

Example 7.2 illustrates that the value of the sample mean will vary from sample to sample. Therefore, the sample mean is a variable and we can make a histogram of all those sample means. This histogram is a visual way to see the **sampling distribution of the sample mean**.

Solution: To obtain the histogram of the sampling distribution of the sample means, enter the sample means in L_1.

1. Be sure that you have no graphs turned on in the Y= window.

2. Press 2nd STAT PLOT to begin setting up the STAT PLOT window. Your screen should look something like Figure 7.4.

3. Choose which of the 3 plots you want to use and make sure that the others are turned off. Select a plot by pressing its number or arrow to it and pressing ENTER. For this example, we will use Plot 1.

4. Using your arrow keys, highlight ON with the cursor and press ENTER.

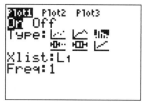

Figure 7.4   Figure 7.5

5. Arrow down to Type and highlight ⼤, the histogram icon, arrow to the right and pressing ENTER.

6. Arrow to Xlist and press 2nd L1 to enter the list as L_1.

7. Arrow to frequency and set it as 1. Note: The TI-83/84 Plus will automatically go into alpha mode when it reaches the list and frequency entries. To return to normal mode, press ALPHA. Your screen should now appear as in Figure 7.5.

Now we must set the window. The window tells the TI-83/84 Plus which portion of the graph you wish to view. It is important to reset the window for every new graph.

8. Press WINDOW.

9. For a histogram, the Xmin is the lowest value in your first class. Here that is 75.

10. The Xmax is the first value beyond your last class. For our example, that is 85.

11. The Xscl is your class width, 11 for this data set.

12. Set the Ymin at 0.

13. The Ymax is set just beyond your highest class frequency. If this is not known, estimate it. Here, a Ymax of 2 will do.

14. Your Yscl should be set based on how far apart your Ymin and Ymax are. Here a Yscl of 1 is sufficient.

The Xres key sets the pixel resolution for function graphs only. For most of our graphs, its setting will not matter. Your window screen should appear as in Figure 7.6. Note: Although the TI-83/84 Plus has a ZoomStat feature it is generally not advisable to use it for graphing histograms.

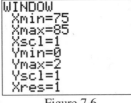

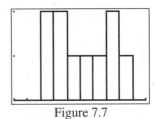

Figure 7.6   Figure 7.7

15. Once the plot and window screens are set, press GRAPH. The TI-83/84 Plus will display your histogram. See Figure 7.7.

The histogram in Figure 7.7 is a visual way to see the **sampling distribution of the sample mean**. In other words, the histogram shows the set of all possible sample means for samples of size 2.

7.1 Practice Problems

Problem 7.17   World's Richest: Each year, *Forbes* magazine publishes a list of the world's richest people. In 2009, the six richest people, their citizenship, and their wealth (to the nearest billion dollars) are as shown Table 7.4. Consider these six people as the population of interest.

Table 7.4

Name	Citizenship	Wealth ($ billion)
William Gates III (G)	United States	40
Warren Buffett (B)	United States	38
Carlos Slim Helu (H)	Mexico	35
Lawrence Ellison (E)	United States	23
Ingvar Kamprad (K)	Sweden	22
Karl Albrecht (A)	Germany	22

a. Calculate the mean wealth, μ, of the six people.
b. For samples of size 2, construct a table similar to Table 7.1. (There are 15 possible samples of size 2.)
c. Construct a histogram for the sampling distribution of the sample mean for samples of size 2.
d. For a random sample of size 2, what is the probability that the sample mean will equal the population mean?
e. For a random sample of size 2, determine the probability that the mean wealth of the two people obtained will be within 2 (i.e., $2 billion) of the population mean. Interpret your result in terms of percentages.

Problem 7.18   Repeat parts (b)–(e) of Problem 7.17 for samples of size 1.

Problem 7.19   Repeat parts (b)–(e) of Problem 7.17 for samples of size 3. (There are 20 possible samples.)

Problem 7.20   Repeat parts (b)–(e) of Problem 7.17 for samples of size 4. (There are 15 possible samples.)

Problem 7.21   Repeat parts (b)–(e) of Exercise 7.17 for samples of size 5. (There are six possible samples.)

Problem 7.22   Repeat parts (b)–(e) of Exercise 7.17 for samples of size 6. What is the relationship between the only possible sample here and the population?

Problem 7.23   Explain what the histograms in part (c) of Problems 7.17–7.22 illustrate about the impact of increasing sample size on sampling error.

LESSON 7.2 THE MEAN AND STANDARD DEVIATION OF THE SAMPLE MEAN

The value of the sample mean will vary from sample to sample. Therefore, the sample mean is a variable. Just like any variable, we can find the mean and the standard deviation of the sample mean.

Example 7.4 Heights of Starting Players: Find the mean of the sample means for all possible samples of size 2 given in Table 7.2. Compare the mean of the sample means to the population mean.

To calculate the mean of the sample means, μ. Enter the data from the last column of Table 7.2 into List

1. Press [STAT].

2. Arrow over to CALC.

3. Press [1] or [ENTER]. **1 - Var Stats** will appear on your home screen. See Figure 7.1.

4. Enter the list where the data are stored, here List 1, [2nd] L1, and press [ENTER]. See Figure 7.2.

5. This screen yields the mean as well as several other statistics. See Figure 7.8.

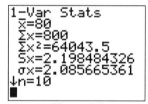

Figure 7.8

We see in Figure 7.8 that the mean of all the sample means is 80. The mean of all possible sample means for all samples of size 2 is equal to the population mean. This example illustrates the following formula:

Formula 7.1: For samples of size n, the mean of the variable $\overline{x}$ equals the mean of the variable under consideration. In symbols, $\mu_{\overline{x}} = \mu$.

Example 7.5 Heights of Starting Players: Find the standard deviation of the sample means for all possible samples of size 2 given in Table 7.2.

Solution: From Figure 7.8, we find that the standard deviation of all possible sample means is 2.19848. This example illustrates the following formula:

Formula 7.2: For samples of size n, the standard deviation of the $\overline{x}$ equals the standard deviation of the variable under consideration divided by the square root of the sample size. In symbols, $\sigma_{\overline{x}} = \sigma/\sqrt{n}$.

LESSON 7.3   THE SAMPLING DISTRIBUTION OF THE MEAN

KEY FACT 7.2: Suppose a random variable x of a population is normally distributed with a mean μ and a standard deviation σ. Then, although it is not obvious, for samples of size n, the random variable $\overline{x}$ is also normally distributed and has mean μ and standard deviation $\sigma/\sqrt{n}$.

An intuitive interpretation of Key Fact 7.2 is as follows. If the sample mean $\overline{x}$ is calculated for every possible sample of size n from a normally distributed population, then the histogram of all those $\overline{x}$ values will be bell-shaped, and have the same mean as the population but with a smaller standard deviation given by the formula $\sigma/\sqrt{n}$.

We can simulate Key Fact 7.2 using the TI-83/84 Plus and generating random samples from a normal distribution and computing the means of the generated samples.

Example 7.8   Intelligence quotients (IQs) measured on the Stanford Revision of the Binet-Simon Intelligence Scale are known to be normally distributed with a mean of 100 and a standard deviation of 16. Use the TI-83/84 Plus to do the following.

a. Simulate 500 samples of four IQs each.[†]

b. Determine the mean of each of the 500 samples.

c. Obtain a histogram of the 500 sample means.

Solution: The solution to this example can be done using the program CH7SIMUL contained in the WeissStats CD and printed in the appendix. Below is an outline of what the program does and then instructions on how to run the program.

The program CH7SIMUL generates a sample and stores it in List 2. Then it finds the mean of that sample and stores it into List 1. The program repeats this step 500 times, storing each mean in the next entry of List 1. When it is finished finding the means, it then creates a histogram of the means. This program uses the ZoomStat feature to set the window for the graph because we are not concerned with setting a particular class width. However, after the program is run the sample mean values will remain in List 1 if you choose to change the window and regraph the histogram.

To run the program:

1. With the program entered in the calculator, from the home screen, press [PRGM].

2. Arrow down the list to the name of the program CH7SIMUL and press [ENTER].

3. prgmCH7SIMUL should appear on your screen. Press [ENTER].

4. The program will ask you for the number of samples. For our example, enter 500 and press [ENTER].

5. The program will ask you for the sample size. For our example, enter 4 and press [ENTER].

6. The program will ask you for the mean. For our example, enter 100 and press [ENTER].

[†] We have chosen to use only 500 samples rather than the 1000 that the textbook uses because of the limits on memory that the TI-83/84 Plus has. Also, this simulation will take about 5 minutes to run on the TI-83/84 Plus when using the program.

7. The program will ask you for the standard deviation. For our example, enter 16 and press ENTER.

8. The program will take awhile to run (approximately 5 min.) During this time, there will be a small moving line at the top right hand corner of your screen. When the program is finished, your histogram will appear on your screen.

Because this simulation involves random sampling, results will differ each time you run the program. However, a typical histogram might look like Figure 7.9.

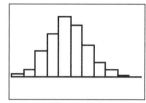

Figure 7.9

KEY FACT 7.3 THE CENTRAL LIMIT THEOREM (CLT): For a relatively large sample size, the variable $\bar{x}$ is approximately normally distributed, regardless of the distribution of the variable under consideration. The approximation becomes better with increasing sample size.

When the population being sampled is normally distributed, then the sampling distribution of the sample mean is normally distributed (Key Fact 7.2). Remarkably, when the sample size is relatively large, the sampling distribution of the sample mean is approximately normally distributed even if the population being sample is not normally distributed. The more non-normal (e.g., more skewed) the population distribution is, the larger the sample size must be for the normal curve to approximate the distribution of $\bar{x}$. A good rule of thumb is to use sample sizes of 30 or more ($n \geq 30$).

7.3 Practice Problems

Problem 7.69   In 1905, R. Pearl published the article "Biometrical Studies on Man. I. Variation and Correlation in Brain Weight" (*Biometrika*, Vol. 4, pp. 13-104). According to the study, brain weights of Swedish mean are normally distributed with a mean of 1.40 kg and a standard deviation of 0.11 kg. Use the TI-83/84 Plus to do the following:

a. Determine the sampling distribution of the sample mean for samples of size 3. Interpret your answer in terms of the distribution of all possible sample mean brain weights for samples of three Swedish men.
b. Repeat part (a) for samples of size 12.
c. Determine the percentage of all samples of three Swedish men that have mean brain weights within 0.1 kg of the population mean brain weight of 1.40 kg. Interpret your answer in terms of sampling error.
d. Repeat part (c) for samples of size 12.

Problem 7.5   As reported in *Runner's World* magazine, the times of the finishers in the New York City 10 km run are normally distributed with a mean of 61 minutes and a standard deviation of 9 minutes. Use the TI-83/84 Plus to do the following.

a. Find the sampling distribution of the sample mean for samples of size 4.
b. Repeat part (a) for samples of size 9.
c. Obtain the percentage of all samples of four finishers that have mean finishing times within 5 minutes of the population mean finishing time of 61 minutes. Interpret your answer in terms of sampling error.
d. Repeat part (c) for samples of size 9.

Problem 7.77   Poverty and Dietary Calcium: Calcium is the most abundant mineral in the human body and has several important functions. Most body calcium is stored in the bones and teeth, where it functions to support their structure. Recommendations for calcium are provided in *Dietary Reference Intakes*, developed by the Institute of Medicine of the National Academy of Sciences. The recommended adequate intake (RAI) of calcium for adults (ages 19–50) is 1000 milligrams (mg) per day. If adults with incomes below the poverty level have a mean calcium intake equal to the RAI, what percentage of all samples of 18 such adults have mean calcium intakes of at most 947.4 mg? Assume that $\sigma = 188$ mg. State any assumptions that you are making in solving this problem.

Problem 7.84   Gestation Periods of Humans: For humans, gestation periods are normally distributed with a mean of 266 days and a standard deviation of 16 days. Suppose that you observe the gestation periods for a sample of nine humans.

a. Simulate 500 samples of nine human gestation periods each.
b. Find the sample mean of each of the 500 samples.
c. Obtain the mean, the standard deviation, and a histogram of the 500 sample means.
d. Theoretically, what are the mean, standard deviation, and distribution of all possible sample means for samples of size 9?
e. Compare your results from parts (c) and (d).

CHAPTER 8
CONFIDENCE INTERVALS FOR ONE POPULATION MEAN

LESSON 8.1 ESTIMATING A POPULATION MEAN

Definition 8.1: A **point estimate** of a parameter is the value of a statistic used to estimate the parameter.

We will use the sample mean as a point estimate of the population mean.

Example 8.1   The U.S. Census Bureau publishes annual price figures for new mobile homes in *Manufactured Housing Statistics*. The figures are obtained from sampling, not from a census. A simple random sample of 36 new mobile homes yielded the prices, in thousands of dollars, shown in Table 8.1. Use the data to find a point estimate of the population mean price, μ, of all new mobile homes

Table 8.1	67.8	68.4	59.2	56.9	63.9	62.2	55.6	72.9	62.6
	67.1	73.4	63.7	57.7	66.7	61.7	55.5	49.3	72.9
	49.9	56.5	71.2	59.1	64.3	64.0	55.9	51.3	53.7
	56.0	76.7	76.8	60.6	74.5	57.9	70.4	63.8	77.9

Solution: We can estimate the population mean price, μ, of all new mobile homes by finding the sample mean price, $\overline{x}$, of the 36 new mobile homes sampled.

We assume that the data are in List 1.

1. Press $\boxed{\text{STAT}}$.

2. Arrow over to CALC.

3. Press $\boxed{1}$ or $\boxed{\text{ENTER}}$. **1 - Var Stats** will appear on your home screen.

4. Enter the list where the data are stored, here List 1, $\boxed{\text{2nd}}$ L1, and press $\boxed{\text{ENTER}}$. See Figure 8.1. This screen yields several summary statistics including the sample mean. We see that the sample mean is 63.28 (rounded to 2 decimal places).

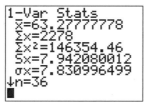

Figure 8.1

Based on the sample data, we estimate the mean price, μ, of all new mobile homes to be approximately $63.28 thousand, that is, $63,280. An estimate of this kind is called a **point estimate** for μ because it consists of a single number, or point.

LESSON 8.2 CONFIDENCE INTERVALS FOR ONE POPULATION MEAN WHEN σ IS KNOWN

Procedure 8.1 of the textbook outlines the steps for determining a **one-sample z-interval** for the population mean.

The assumptions for using this interval are:

1. Simple random sample.

2. Normal population or large sample.

3. σ known.

Key Fact 8.1 of Section 8.2 of the textbook discusses when it is appropriate to use the z-interval procedure. Verification of the assumptions involves constructing normal probability plots, boxplots, and histograms. All these graphs can be done on the TI-83/84 Plus.

Example 8.4 The U.S. Bureau of Labor Statistics collects information on the ages of people in the civilian labor force and publishes the results in *Employment and Earnings*. Fifty people in the civilian labor force are randomly selected; their ages are displayed in Table 8.2. Find a 95% confidence interval for the mean age, μ, of all people in the civilian labor force. Assume σ = 12.1 years.

Table 8.2

22	58	40	42	43	32	34	45	38	19
33	16	49	29	30	43	37	19	21	62
60	41	28	35	37	51	37	65	57	26
27	31	33	24	34	28	39	43	26	38
42	40	31	34	38	35	29	33	32	33

Solution: Using the TI-83/84 Plus we constructed a normal probability plot, histogram, and boxplot for these data. The boxplot indicated potential outliers, but after looking at the other graphs, we concluded that in fact, there are no outliers. The sample size is large (n = 50) and σ is known (12.1 years). Therefore we can use the z-interval procedure.

There are two ways to construct a z-interval using the TI-83/84 Plus. The first is useful if you have the data values but have not computed the mean. We will use this method first.

1. With the data in List 1, press $\boxed{\text{STAT}}$ and arrow over to the TESTS menu.

2. Number 7 is the ZInterval so either press $\boxed{7}$ or arrow down to 7 and press $\boxed{\text{ENTER}}$.

3. Highlight Data and press $\boxed{\text{ENTER}}$.

4. Enter your σ, 12.1, and press $\boxed{\text{ENTER}}$.

5. Set the List as List 1 by pressing $\boxed{\text{2nd}}$ L1.

6. Set the Freq: as 1. Note that the calculator goes into ALPHA mode when it reaches the Freq: line. To change to normal mode, press $\boxed{\text{ALPHA}}$.

7. Set the C-Level as 0.95. Your screen should appear as in Figure 8.2.

8. Highlight Calculate and press $\boxed{\text{ENTER}}$. Your answer will be displayed as in Figure 8.3.

Our confidence interval is from 33.026 to 39.734 or rounded to one decimal place, from 33.0 to 39.7. Note that the textbook gives the answer as 33.0 to 39.8. This difference is due to the fact that they round the mean to 36.4 whereas the TI-83/84 Plus used 36.38 to calculate the interval. Also, the TI-83/84 Plus does not round the z-score when doing the computation.

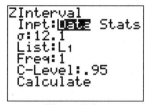

Figure 8.2

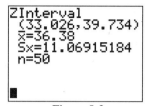

Figure 8.3

The second method of computing the interval is useful if you do not have the data values but have the sample mean and sample size.

1. Press STAT and arrow over to the TESTS menu.

2. Number 7 is the ZInterval so either press 7 or arrow down to 7 and press ENTER.

3. Highlight Stats and press ENTER.

4. Enter your σ, 12.1, and press ENTER.

5. Enter your $\overline{x}$ as 36.4 and press ENTER.

6. Enter your n of 50 and press ENTER.

7. Enter the C-Level as 0.95 and press ENTER. Your screen should appear as in Figure 8.4.

8. Highlight Calculate and press ENTER. Your answer will be displayed as in Figure 8.5.

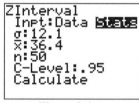

Figure 8.4

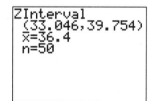

Figure 8.5

Notice, our results are slightly different than when we used the first method. This time our results match the textbook when rounded to one decimal place.

We can be 95% confident that the mean age, μ, of all people in the civilian labor force is somewhere between 33.0 years and 39.8 years.

8.2 Practice Problems

Problem 8.31 Data on investments in high-tech industry by venture capitalists are compiled by VentureOne Corporation and published in *America's Network Telecom Investor Supplement*. A random sample of 18 venture capital investments in the fiber optics business sector yielded the data, in millions dollars, in Table 8.3. Determine and interpret a 95% confidence interval for the mean amount, μ, of all venture-capital investments in the fiber optics business sector. Assume that the population standard deviation is $2.04 million.

Table 8.3	5.60	6.27	5.96	10.51	2.04	5.48
	5.74	5.58	4.13	8.63	5.95	6.67
	4.21	7.71	9.21	4.98	8.64	6.66

Problem 8.32 Calcium is the most abundant mineral in the human body and has several important functions. Most body calcium is stored in the bones and teeth, where it functions to support their structure. Recommendations for calcium are provided in *Dietary Reference Intakes*, developed by the Institute of Medicine of the National Academy of Sciences. The recommended adequate intake (RAI) of calcium for adults (ages 19–50) is 1000 milligrams (mg) per day. A simple random sample of *18* adults with incomes below the poverty level gave the following daily calcium intakes in Table 8.4. Determine a 95% confidence interval for the mean calcium intake, μ, of all adults with incomes below the poverty level. Assume that the population standard deviation is 188 mg. Interpret your answer.

Table 8.4	886	633	943	847	934	841
	1193	820	774	834	1050	1058
	1192	975	1313	872	1079	809

Problem 8.36 The Rolling Stones, a rock group formed in the 1960s, has toured extensively in support of new albums. *Pollstar* has collected data on the earnings from the Stones' North American tours. For 30 randomly selected Rolling Stones concerts, the mean gross earnings is $2.27 million. Assuming a population standard deviation gross earnings of $0.5 million, obtain and interpret a 99% confidence interval for the mean gross earnings of all Rolling Stones concerts.

Problem 8.42 R. Reifen et al. studied various nutritional measures of Ethiopian school children and published their findings in the paper "Ethiopian-Born and Native Israeli School Children Have Different Growth Patterns" (*Nutrition*, Vol. 19, pp. 427-431). The study, conducted in Azezo, North West Ethiopia, found that malnutrition is prevalent in primary and secondary school children because of economic poverty. The weights in kilograms (kg), of 60 randomly selected male Ethiopian-born school children, ages 12-15 years old, are presented in Table 8.5. Find and interpret a 95% confidence interval for the mean weight of all male Ethiopian-born school children, ages 12-15 years old. Assume that the population standard deviation is 4.5 kg.

Table 8.5	45.7	48.9	53.8	44.7	42.8	50.9
	38.9	45.2	48.1	42.0	48.2	42.1
	52.8	45.9	47.9	46.2	42.9	42.5
	46.5	45.9	56.6	51.8	43.4	43.3
	40.9	44.4	45.2	44.0	47.8	36.3
	39.3	42.0	38.0	37.7	42.5	44.8
	48.3	47.4	45.4	48.5	43.5	41.3
	49.2	45.2	46.8	46.6	41.1	45.5
	47.5	48.4	41.8	49.1	49.5	41.5
	46.3	48.6	51.4	39.0	38.8	47.2

LESSON 8.3   MARGIN OF ERROR

Definition 8.3: The **margin of error** for the estimate of μ is given by

$$E = z_{\alpha/2} \frac{\sigma}{\sqrt{n}}$$

KEY FACT 8.4: The length of a confidence interval for a population mean, μ, and therefore the precision with which $\overline{x}$ estimates μ is determined by the margin of error, E. For a fixed confidence level, increasing the sample size improves the precision and vice versa.

Sample Size for Estimating μ

We have already seen in Key Fact 7.1 that the larger the sample size, the smaller the **sampling error** tends to be. The margin of error and the confidence level of a confidence interval are often specified in advance. We must then determine the required **sample size** to meet these specifications. The formula for the sample size required for a $(1 - \alpha)$-level confidence interval for μ with a specified margin of error, E, is given by the formula

$$n = \left(\frac{z_{\alpha/2} \cdot \sigma}{E} \right)^2 .$$

LESSON 8.4   CONFIDENCE INTERVALS FOR ONE POPULATION MEAN WHEN σ IS UNKNOWN

Finding the Area Under the T-Curve

The TI-83/84 Plus has a built-in function to find the area under a t-curve with specified degrees of freedom to the left of a t-value.

Example   For a t-curve with df = 12, find the area under the curve to the left of 1.643.

Solution:

     1. Press [2nd] DISTR to access the distribution menu.

     2. Press [5] or arrow down to **5:tcdf(** and press [ENTER].

     3. The format of this command is **tcdf(** lowerbound, upperbound,df). For our problem we need to find the area from -∞ to 1.643. To estimate -∞ for the calculator, we will use -1E99 as we did when computing normal distribution areas. Recall that we access the E by pressing [2nd] EE. For our problem our command is therefore **tcdf(**-1E99, 1.643,12). The result is shown in Figure 8.6.

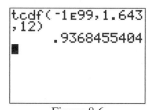

Figure 8.6

If you are looking for the area to the right of a t-value, you will need to estimate ∞ as 1E99. Keep in mind that the t-curve is symmetric about 0 and that the total area under the curve is equal to 1.

Finding the t-value Having a Specified Area to the Right

The TI-84 Plus v2.4 has a built-in InvT function similar to InvNorm. The TI-83 and earlier versions of the TI84 do not. However, the program INVT contained in the WeissStats CD can be used to accomplish the same thing. Both methods will be demonstrated.

Example 8.8 For a t-curve with $df = 13$, determine $t_{0.05}$; that is find the t-value having area 0.05 to its right.

Solution:

Using the TI-84 Plus v 2.4

1. Press [2nd] DISTR to access the distribution menu.

2. Arrow down to **5:tcdf(** and press [ENTER].

3. The format of this command is **invT(** area to the left,df). For our problem we need to find the t-value having an area 0.05 to the right or 1 - 0.05 = 0.95 area to the left with 13 *df*. Therefore, the command is invT(1-0.05,13). The result is shown in Figure 8.6.

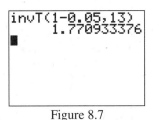

Figure 8.7

Using the TI-83 Plus

1. Press [PRGM], arrow down to INVT and press [ENTER]. prgmINVT will appear on your home screen.

2. Press [ENTER] to run the program.

3. Follow the prompts and enter 13 for the DF, 2 for area to the right, and 0.05 for the area. The calculator will take a little while to do the computation. See Figure 8.8 for the final result.

The T-value with 13 degrees of freedom having area 0.05 to its right is 1.771 when rounded to three decimal places.

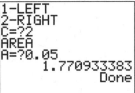

Figure 8.8

There is a slight difference in the values the TI-83 Plus and the TI-84 Plus compute because of the methods used in the computations. However, the difference disappears when the values are rounded to three decimal places and should not impact your answers.

8.4 Practice Problems

Problem 8.81 For a t-curve with df = 6, find each of the following t-values
a) $t_{0.10}$ b) $t_{0.025}$ c) $t_{0.01}$

Problem 8.82 For a t-curve with df = 17, find each of the following t-values
a) $t_{0.05}$ b) $t_{0.025}$ c) $t_{0.005}$

Confidence Intervals for One Population Mean When σ Is Unknown

Procedure 8.2 outlines the steps for determining a **one-sample t-interval.**

The assumptions for this interval are

1. Simple random sample

2. Normal population or large sample

3. σ unknown

The guidelines for using a t-interval are the same as those for the z-interval.

Example 8.9   The U.S. Federal Bureau of Investigation (FBI) compiles data on robbery and property crimes and publishes the information in *Population-at-Risk Rates and Selected Crime Indicators*. A sample of a recent year's pick-pocket offenses yields the values lost shown in Table 8.6. Use the data to obtain a 95% confidence interval for the mean value lost, μ, of all last year's pick-pocket offenses.

Table 8.6	447	207	627	430	883
	313	844	253	397	214
	217	768	1064	26	587
	833	277	805	653	549
	649	554	570	223	443

Solution: Since the sample size is 25, we must first consider the questions of normality and outliers. To do that we construct a normal probability plot of the data as shown in Figure 8. The window for this plot is given in Figure 8.10.

Figure 8.9

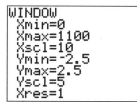

Figure 8.10

The normal plot shows no outliers and falls roughly in a straight line. Therefore, we can use the t-interval procedure.

Like the z-interval, there are two ways to do a t-interval on the TI-83/84 Plus. The first is best used when you have the data values, but do not have the sample mean and sample standard deviation.

1. With the data in List 1, press STAT and arrow over to the TESTS menu.

2. Number 8 is the TInterval so either press 8 or arrow down to 8 and press ENTER.

3. Highlight Data and press ENTER.

4. Press 2nd L1 to enter List 1.

5. Set the Freq: as 1.

6. Set the C-Level as 0.95. Your screen should appear as in Figure 8.11.

7. Highlight Calculate and press ENTER. Your answer will be displayed as in Figure 8.12.

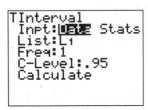   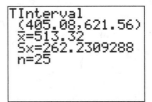

| Figure 8.11 | Figure 8.12 |

Once again, due to the rounding of the t-value and the standard deviation, the textbook answers may vary slightly from the TI-83/84 Plus's answers. Here the confidence interval is 405.08 to 621.56. We can be 95% confident that the mean value lost, μ, of all last year's pick-pocket offenses is somewhere between $405.08 and $621.56.

The second method of computing the interval is useful if you do not have the data values but have the sample mean, sample standard deviation and sample size.

1. Press STAT and arrow over to the TESTS menu.

2. Number 8 is the TInterval so either press 8 or arrow down to 8 and press ENTER.

3. Highlight Stats and press ENTER.

4. Enter your sample mean as 513.32.

5. Enter your sample standard deviation as 262.23.

6. Enter your sample size as 25.

7. Enter the C-Level as 0.95. See Figure 8.13.

8. Highlight Calculate and press ENTER. The answer is given in Figure 8.14.

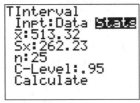   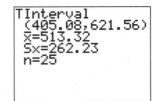

| Figure 8.13 | Figure 8.14 |

Once again, the answer is slightly different from the textbook because of the rounding of the t-value.

8.4 Practice Problems (continued)

Problem 8.96 Taking the family to an amusement park has become increasingly costly according to the industry publication *Amusement Business*, which provides figures on the cost for a family of four to spend the day at one of America's amusement parks. A random sample of 25 families of four that attended amusement parks yielded the costs in Table 8.7, rounded to the nearest dollar. Find and interpret a 95% confidence interval for the mean cost of a family of four to spend the day at an American amusement park. Preliminary data analyses indicate that it is reasonable to apply the t-procedure.

Table 8.7	122	166	171	148	135
	173	137	163	119	144
	164	153	162	140	142
	158	130	167	173	186
	92	170	126	163	172

Problem 8.103 The subterranean coruro (*Spalacopus cyanus*) is a social rodent that lives in large colonies in underground burrows that can reach lengths of up to 600 meters. Zoologists Sabine Begall and Milton H. Gallardo studied the characteristics of the burrow systems of the subterranean coruro in central Chile and published their findings in the *Journal of Zoology, London*(Vol. 251, pp. 53-60). A sample of 51 burrows had the depths in Table 8.8, in centimeters. Find and interpret a 90% confidence interval for the mean depth of all subterranean coruro burrows. Preliminary data analyses indicate that it is reasonable to apply the t-procedure.

Table 8.8	15.1	16.0	18.3	18.8	13.9	15.8	14.2
	12.3	11.8	12.1	17.9	16.6	16.5	16.0
	12.8	14.7	15.9	13.9	17.2	12.2	18.2
	16.9	13.3	14.4	15.0	12.1	11.0	16.7
	17.4	8.2	19.3	17.4	15.3	15.6	19.7
	14.5	12.5	12.8	13.3	16.8	17.5	14.0
	14.9	16.7	12.0	15.0	16.2	9.7	15.4
	18.9	14.9					

Problem 8.106 A city planner working on bikeways needs information about local bicycle commuters. She designs a questionnaire. One of the questions asks how many minutes it takes the rider to pedal from home to his or her destination. A sample of local bicycle commuters yields the times in Table 8.9. Find and interpret a 90% confidence interval for the mean commuting time of all local bicycle commuters in the city.

Table 8.9	22	19	24	31	29	29
	21	15	27	23	37	31
	30	26	16	26	12	
	23	48	22	29	28	

Review
Problem #18 A paper by Cho et al. in the May 2000 issue of *The Journal of Pediatrics* (Vol. 136(5), pp. 587-592) presented the results of research on various characteristics in children of diabetic mothers. Past studies have shown that maternal diabetes results in obesity, blood pressure, and glucose tolerance complications in the offspring. Table 8.10 contains the arterial blood pressures, in millimeters of mercury, for a random sample of 16 children of diabetic mothers. Find a 95% confidence interval for the mean arterial blood pressure for all children of diabetic mothers.

Table 8.10	81.6	84.1	87.6	82.8	84.6	104.9	90.8	94.0
	82.0	88.9	86.7	96.4	69.4	78.9	75.2	91.0

CHAPTER 9
HYPOTHESIS TESTS FOR ONE POPULATION MEAN

LESSON 9.1   THE NATURE OF HYPOTHESIS TESTING

Before we use the TI 83/84 Plus to carry out any hypothesis tests concerning one population mean, μ we will begin with a general discussion of **hypothesis testing**. We will also define some terms used in hypothesis testing.

A hypothesis test involves two hypotheses– the **null hypothesis** (H_0) and the **alternative hypothesis** (H_a). The goal in a hypothesis testing is to decide which of the two hypotheses is true. That is, to decide whether or not to reject the null hypothesis in favor of the alternative hypothesis. This decision is based on a random sample selected from the population.

The null hypothesis for a hypothesis test concerning a single population mean, μ, usually states that the mean is equal to one specific number. For example, the null hypothesis is of the form

$$H_0: \quad \mu = \mu_0$$

where μ_0 is the specified number.

On the other hand, the alternative hypothesis can be any *one* of the three different possible forms:

$$H_a: \quad \mu \neq \mu_0, \quad H_a: \quad \mu < \mu_0, \quad \text{or} \quad H_a: \quad \mu > \mu_0.$$

The first hypothesis test is a **two-tailed test**; the second case is a **left-tailed test**; and the third case is a **right-tailed test**. Left-tailed and right-tailed tests are also called **one-tailed tests**.

Recall the goal of hypothesis testing is to decide which of the two hypotheses the random sample from the population provides evidence to support. When making decisions based on a random sample there is always the possibility of making an error. In fact, two different errors can be made. A **Type I error** is made if the null hypothesis is rejected when in fact it is true. A **Type II error** is made when the null hypothesis is not rejected when in fact the null hypothesis is false.

The probability of a Type I error is called the **significance level** of the hypothesis and denoted by α. The significance level is usually taken to be between 0.01 and 0.10.

LESSON 9.2   CRITICAL VALUE APPROACH TO HYPOTHESIS TESTING

The set of values for the test statistic that lead to the null hypothesis being rejected is called the **rejection region**. A choice between the null and alternative hypotheses can be made by checking if the test statistic falls into the **rejection region** or into the **nonrejection region** for the test.

The value of the test statistic that separates the rejection and nonrejection region is called the **critical value**.

The textbook discusses several different hypothesis testing procedures for testing a claim about a population mean. The first procedure your textbook discusses is used when the population standard deviation, σ is known. This hypothesis test is called the **one-mean z-test**. In this situation, this decision of whether or not to reject the null hypothesis is based on the test statistic $z = \dfrac{\overline{x} - \mu_0}{\sigma / \sqrt{n}}$ which tells how many standard deviations the sample mean is

from the null hypothesis population mean of μ_0. In this procedure, the critical value would be a z-score.

The second procedure the textbook discusses is used when the population standard deviation σ is unknown. This hypothesis test is called the **one-mean t-test** because it is based on the Student t-distribution. In this situation, this decision of whether or not to reject the null hypothesis is based on the test statistic $t = \dfrac{\bar{x} - \mu_0}{s/\sqrt{n}}$, which tells how many standard deviations the sample mean is from the null hypothesis population mean of μ_0. In this procedure, the critical value would be a t-score.

LESSON 9.3 *P*-VALUE APPROACH TO HYPOTHESIS TESTING

The **p-value** for a hypothesis test is the smallest significance level at which the null hypothesis can be rejected based on the test statistic calculated from the random sample that has been selected.

The decision criterion for a hypothesis test using the *p*-value is to *reject the null hypothesis* if the *p*-value is less than or equal to the significance level α. Otherwise, do not reject the null hypothesis.

The *p*-value of a two-tailed hypothesis test will be twice the *p*-value of a one-tailed hypothesis test.

LESSON 9.4 HYPOTHESIS TESTS FOR ONE POPULATION MEAN WHEN σ IS KNOWN

Procedure 9.1 of the textbook outlines the steps for performing a one-sample z-test using either the *critical value* approach or the *p*-value approach. The TI-83/84 Plus can be used to calculate the test statistic and the *p*-value.

Recall the assumptions for this test are the same as for the one-sample z-confidence interval. They are

    1. Simple random sample

    2. Normal population or large sample

    3. σ is known.

The preliminary data analyses steps on constructing normal probability plots, boxplots, and histograms can be done on the TI-83/84 Plus.

Example 9.12   Calcium is the most abundant mineral in the body and also one of the most important. It works with phosphorus to build and maintain bones and teeth. According to the Food and Nutrition Board of the National Academy of Sciences, the recommended daily allowance (RDA) of calcium for adults is 1000 milligrams (mg). A random sample of 18 people with incomes below the poverty level gives the daily calcium intakes shown in Table 9.1. At the 5% significance level, do the data provide sufficient evidence to conclude that the mean calcium intake of all people with incomes below the poverty level is less than the RDA of 1000 mg? Assume σ = 188 mg.

Table 9.1	886	4633	943	847	934	841
	1193	820	774	834	1050	1058
	1192	975	1313	872	1079	809

Solution: Let μ denote the mean calcium intake (per day) of all people with incomes below the poverty level.

Step 1: State the null and alternative hypotheses.

H_o: μ = 1000 mg (mean calcium intake is not less than the RDA)

H_a: μ < 1000 mg (mean calcium intake is less than the RDA).

Step 2: Decide on the significance level.

We are to perform the hypothesis test at the 5% significance level; so $\alpha = 0.05$.

Step 3: Compute the value of the test statistic.

Because we know σ, we can use the one-sample z-test. There are two ways to perform a one-sample z-test using the TI-83. The first is useful if you have the data values but have not computed the sample mean. We will use this method first.

1. With the data in List 1, press [STAT] and arrow over to the TEST menu.

2. Number 1 is the Z-Test so press [1] or press [ENTER].

3. Highlight Data and press [ENTER].

4. Enter μ_0 as 1000 and press [ENTER].

5. Enter σ as 188 and press [ENTER].

6. Enter List as List 1 by pressing [2nd] L1.

7. Enter Freq: as 1 (Recall the TI-83/84 Plus goes into Alpha mode for Frequencies so press [ALPHA] [1].)

8. Highlight the appropriate alternative hypothesis and press [ENTER]. We will highlight $< \mu_o$ for our example. Your screen should appear as in Figure 9.1.

The TI-83/84 Plus has two displays for the answer. We will demonstrate both.

9. Highlight Calculate and press [ENTER]. Your screen will appear as in Figure 9.2. If you highlight Draw and press [ENTER], your screen will appear as in Figure 9.3.

Figure 9.1

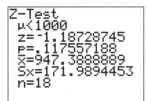
Figure 9.2

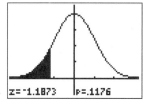
Figure 9.3

The second method of computing the test statistic and the p-value is useful if you do not have the data values but have the sample mean, sample standard deviation and sample size.

1. Press [STAT] and arrow over to the TEST menu.

2. Number 1 is the Z-Test so press [1] or press [ENTER].

3. Highlight Stats and press [ENTER].

4. Enter μ_0 as 1000 and press ENTER.

5. Enter σ as 188 and press ENTER.

6. Enter the sample mean, 947.4 and press ENTER.

7. Enter the sample size, 18, and press ENTER.

8. Highlight the appropriate alternative hypothesis and press ENTER. We will highlight $< \mu_0$ for our example. Your screen should appear as in Figure 9.4.

The TI-83/84 Plus has two displays for the answer. We will demonstrate both.

9. Highlight Calculate and press ENTER. Your screen will appear as in Figure 9.5. If you highlight Draw and press ENTER, your screen will appear as in Figure 9.6.

Figure 9.4

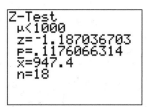
Figure 9.5

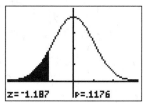
Figure 9.6

Note that both answer screens for both methods display the test statistic and the p-value. Here the test statistic is given as z = -1.19 when rounded to two decimal places.

Step 4: Obtain the *p*-value.
The *p*-value is 0.1176 when rounded to four decimal places.

When using the two different methods, your answers for the test statistic and p-value may vary slightly and the values may differ slightly from the answers in the textbook because the TI-83/84 Plus does not round the mean when computing the test statistic and it does not round the test statistic to compute the p-value. However, your final conclusion will always be the same.

Step 5: If *p*-value $\leq \alpha$ reject H_0; otherwise do not reject H_0.
Our *p-value* is 0.1176 which exceeds the specified significance level of 0.05. Therefore we do not reject the null hypothesis.

Step 6: Interpret the results of the hypothesis test.
At the 5% significance level, the sample of 18 calcium intakes does not provide sufficient evidence to conclude that the mean calcium intake, μ, of all people with incomes below the poverty level is less than the RDA of 1000 mg.

9.4 Practice Problems

Problem 9.73   Cadmium, a heavy metal, is toxic to animals. Mushrooms, however, are able to absorb and accumulate cadmium at high concentrations. The Czech and Slovak governments have set a safety limit for cadmium in dry vegetables at 0.5 parts per million (ppm). M. Melgar et al. measured the cadmium levels in a random sample of the edible mushroom *Boletus pinicola* and publishes the results in the *Journal of Environmental Science and Health* (Vol. B33(4), pp. 439-455). The data is in Table 9.2. At the 5% significance level, do the data provide sufficient evidence to conclude that the mean cadmium level in *Boletus pinicola* mushrooms is greater than the government-recommended limit? Assume the population standard deviation of cadmium levels in *Boletus pinicola* mushrooms is 0.37 ppm.

Table 9.2	0.24	0.59	0.62	0.16	0.77	1.33
	0.92	0.19	0.33	0.25	0.59	0.32

Problem 9.74   The R.R. Bowker Company of New York collects information on the retail prices of books and publishes its findings in *The Bowker Annual Library and Book Trade Almanac*. In 2005, the mean retail price of all history books was $57.61. This year's retail prices for 28 randomly selected history books are shown in Table 9.3. At the 10% significance level, do the data provide sufficient evidence to conclude that this year's mean retail price of all history books has changed from the 2005 mean of $57.61? Assume the standard deviation of prices for this year's history books is $8.45.

Table 9.3	59.54	67.70	57.10	46.11	46.86	62.87	66.40
	52.08	37.67	50.47	60.42	38.14	58.21	47.35
	50.45	71.03	48.14	66.18	59.36	41.63	53.66
	49.95	59.08	58.04	46.65	66.76	50.61	66.68

Problem 9.81   A study by researchers at the University of Maryland addressed the question of whether the mean body temperature of humans is 98.6° F. The results of the study by P. Mackowiak, S. Wasserman, and M. Levine appeared in the article "A Critical Appraisal of 98.6° F, the Upper Limit of the Normal Body Temperature and Other Legacies of Carl Reinhold August Wunderlich" (*Journal of the American Medical Association*, Vol. 268, pp. 1578-1580). Among other data, the researchers obtained the body temperatures of 93 healthy humans. At the 1% significance level, do the data provide sufficient evidence to conclude that the mean body temperature of healthy humans differs from 98.6° F? Assume that $\sigma = 0.63$°F and the sample mean is 98.12°F.

LESSON 9.5   HYPOTHESIS TESTS FOR ONE POPULATION MEAN
WHEN σ IS UNKNOWN

Usually the population standard deviation is not known. Procedure 9.3 of Section 9.6 outlines the steps for performing a one-sample t-test. The TI-83/84 Plus can be used to calculate the test statistic and the *p*-value.

Recall the assumptions for this test are the same as for the one-sample t-confidence interval. They are

1. Simple random sample

2. Normal population or large sample

3. σ unknown.

The preliminary data analyses steps on constructing normal probability plots, boxplots, and histograms can be done on the TI-83/84 Plus.

Example 9.16   Acid rain, from the burning of fossil fuels, has caused many of the lakes around the world to become acidic. The biology in these lakes often collapses because of the rapid and unfavorable changes in water chemistry. A lake is classified as non acidic if it has a pH greater than 6. Aldo Marchetto and Andrea Lami measured the pH of high mountain lakes in the Southern Alps and reported their findings in the paper "Reconstruction of pH by Chrysophycean Scales in Some Lakes of the Southern Alps" (*Hydrobiologia*, Vol. 274, pp. 83-90). Table 9.4 provides the pH levels obtained by the researchers for 15 lakes. At the 5% significance level, do the data provide sufficient evidence to conclude that, on the average, high mountain lakes in the Southern Alps are nonacidic?

Table 9.4	7.2	7.3	6.1	6.9	6.6
	7.3	6.3	5.5	6.3	6.5
	5.7	6.9	6.7	7.9	5.8

Solution: Graphical analyses of the data in Table 9.4 reveal no outliers and is quite linear. Therefore, we can use the one-sample t-test.

Step 1: State the null and alternative hypotheses.
Let μ denote the mean pH level of all high mountain lakes in the Southern Alps. Then the null and alternative hypotheses are

H_0: $\mu = 6$ (mean pH in not greater than 6).

H_a: $\mu > 6$ (mean pH level is greater than 6).

Step 2: Decide on the significance level.
We are to perform the hypothesis test at the 5% significance level; so $\alpha = 0.05$.

Step 3: Compute the value of the test statistic.

There are two ways to perform a one-sample t-test using the TI-83. The first is useful if you have the data values but have not computed the sample mean. We will use this method first.

1. Press [STAT] and arrow over to the TEST menu.

2. Number 2 is the T-test so press [2] or arrow down to 2 and press [ENTER].

3. Highlight Data and press [ENTER].

4. Enter μ_0 as 6 and press [ENTER].

5. Enter List as List 1 by pressing [2nd] L1.

6. Enter Freq: as 1. Recall the TI-83/84 Plus goes into Alpha mode for Frequencies so press [ALPHA] [1].

7. Highlight the appropriate alternative hypothesis and press [ENTER]. We will highlight $> \mu_0$ for our example. Your screen should appear as in Figure 9.7.

The TI-83/84 Plus has two displays for the answer. We will demonstrate both.

8. Highlight Calculate and press ENTER. Your screen will appear as in Figure 9.8. If you highlight Draw and press ENTER, your screen will appear as in Figure 9.9.

Figure 9.7

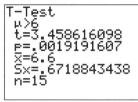
Figure 9.8

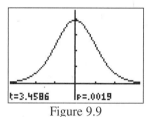
Figure 9.9

The second method of computing the test statistic and the p-value is useful if you do not have the data values but have the sample mean, sample standard deviation and sample size.

1. Press STAT and arrow over to the TEST menu.

2. Number 2 is the T-test so press 2 or arrow down to 2 and press ENTER.

3. Highlight Stats and press ENTER.

4. Enter μ_0 as 6 and press ENTER.

5. Enter the sample mean, 6.6 and press ENTER.

6. Enter the sample standard deviation, 0.672 and press ENTER.

7. Enter the sample size, 15, and press ENTER.

8. Highlight the appropriate alternative hypothesis and press ENTER. We will highlight $> \mu_0$ for our example. Your screen should appear as in Figure 9.10.

The TI-83/84 Plus has two displays for the answer. We will demonstrate both.

9. Highlight Calculate and press ENTER. Your screen will appear as in Figure 9.11. If you highlight Draw and press ENTER, your screen will appear as in Figure 9.12.

Figure 9.10

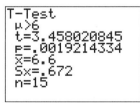
Figure 9.11

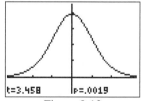
Figure 9.12

Note that both answer screens for both methods display the test statistic and the *p*-value. Here the test statistic is given as t = 3.458 when rounded to three decimal places and the *p*-value is 0.0019 when rounded to 4 decimal places. When using the two different methods, your answers will vary slightly and the values may differ slightly from the answers in the textbook because the TI-83/84 Plus does not round the mean to compute the test statistic. The TI-83/84 Plus also allows you to calculate a more exact value of *p* than the table does.

Step 4: Obtain the *p*-value.
In Figure 9.11, the p-value is 0.0019 when rounded to 4 decimal places.

Step 5: If *p*-value $\leq \alpha$ reject H_0; otherwise do not reject H_0.

Our *p*-value is 0.0019, which is smaller than the specified significance level of 0.05. Therefore we reject the null hypothesis.

Step 6: Interpret the results of the hypothesis test.
At the 5% significance level, the data provide sufficient evidence to conclude that, on the average, high mountain lakes in the Southern Alps are nonacidic.

9.5 Practice Problems

Problem 9.101   According to *Communications Industry Forecast & Report*, published by Veronis Suhler Stevenson, the average person watched 4.55 hours of television per day in 2005. A random sample of 20 people gave the number of hours of television watched per day for last year given in Table 9.5. At the 10% significance level, do the data provide sufficient evidence to conclude that the amount of television watched per day last year by the average person differed from that in 2005? Preliminary data analyses indicate that you can reasonably use a t-test to conduct the hypothesis test required. Preliminary data analyses suggest it is appropriate to use the one-mean t-test.

Table 9.5	1.0	4.6	5.4	3.7	5.2
	1.7	6.1	1.9	7.6	9.1
	6.9	5.5	9.0	3.9	2.5
	2.4	4.7	4.1	3.7	6.2

Problem 9.103 Because many industrial wastes contain nutrients that enhance crop growth, efforts are being made, for environmental purposes, to use such wastes on agricultural soils. Two researchers, Mohammed Ajmal and Ahsan Ullah Khan, reported their findings on experiments with brewery wastes used for agricultural purposes in the article "Effects of Brewery Effluent on Agricultural Soil and Crop Plants" (*Environmental Pollution (Series A)*, 33, pp. 341-351). The researchers studied the physico-chemical properties of effluent from Mohan Meakin Breweries Ltd. (MMBL), Ghazibad, UP, India, and "… its effects on the physico-chemical characteristics of agricultural soil, seed germination pattern, and the growth of two common crop plants." They assessed the impact of using different concentrations of the effluent: 25%, 50%, 75%, and 100%. The data in Table 9.6, based on the results of the study, provide the percentages of limestone in the soil using 100% effluent. Do the data provide sufficient evidence to conclude, at the 1% level of significance, that the mean available limestone in soil treated with 100% MMBL effluent exceeds 2.30%, the percentage ordinarily found? Preliminary data analyses suggest it is appropriate to use the one-mean t-test.

Table 9.6	2.41	2.31	2.54	2.28	2.72
	2.60	2.51	2.51	2.42	2.70

Problem 9.104   According to the document *Consumer Expenditures*, a publication of the U.S. Bureau of Labor Statistics, the average consumer unit spent $1874 on apparel and services in 2006. That same year, 25 consumer units in the Northeast had the annual expenditures, in dollars, on apparel and services listed in Table 9.7. At the 5% significance level, do the data provide sufficient evidence to conclude that the 2006 mean annual expenditure on apparel and services for consumer units in the Northeast differed from the national mean of $1874? Preliminary data analyses suggest it is appropriate to use the one-mean t-test.

Table 9.7

1417	1595	2158	1820	1411
2361	2371	2330	1749	1872
2826	2167	2304	1998	2582
1982	1903	2405	1660	2150
2128	1889	2251	2340	1850

Problem 9.105   The Ankle Brachial Index (ABI) compares the blood pressure of a patient's arm to the blood pressure of the patient's leg. The ABI can be an indicator of different diseases, including arterial diseases. A healthy (or normal) ABI is 0.9 or greater. In a study by M. McDermott et al. titled "Sex differences in Peripheral Arterial Disease: Leg Symptoms and Physical Functioning" (*Journal of the American Geriatrics Society*, Vol. 51, No. 2, pp. 222-228), the researchers obtained the ABI of 187 women with peripheral arterial disease. The results were a mean ABI of 0.64 with a standard deviation of 0.15. At the 5% significance level, do the data provide sufficient evidence to conclude that women with peripheral arterial disease have an unhealthy ABI? Preliminary data analyses suggest it is appropriate to use the one-mean t-test.

Problem 9.107   Cardiovascular Hospitalizations: From the Florida State Center for Health Statistics report, *Women and Cardiovascular Disease Hospitalizations*, we found that, for cardiovascular hospitalizations, the mean age of women is 71.9 years. At one hospital, a random sample of 20 of its female cardiovascular patients had the ages, in years, given in Table 9.8. Decide whether applying the t-test to perform a hypothesis test for the population mean in question appears reasonable. Explain your answers.

Table 9.8

75.9	83.7	87.3	74.5	82.5
78.2	76.1	52.8	56.4	53.8
88.2	78.9	81.7	54.4	52.7
58.9	97.6	65.8	86.4	72.4

Review Problem #38   According to *Food Consumption, Prices, and Expenditures*, published by the U.S. department of Agriculture, the mean consumption of beef per person in 2002 was 64.5 lb. (boneless, trimmed weight). A sample of 40 people taken this year yielded the data in Table 9.9, in pounds, on last year's beef consumptions. At the 5% significance level, do the data provide sufficient evidence to conclude that last year's mean beef consumption is less than the 2002 mean of 64.5 lb.? Preliminary data analyses suggest it is appropriate to use the one-mean t-test.

Table 9.9

77	65	57	54	68	79	56	0
50	49	51	56	56	78	63	72
0	62	74	61	61	60	56	37
76	77	67	67	62	89	56	75
69	73	75	62	8	74	20	47

LESSON 9.6   THE WILCOXON SIGNED-RANK TEST*

Both the z-test and the t-test require the variable under consideration to be approximately normally distributed or the sample should be relatively large and, for small samples, both procedures should be avoided if outliers are present. The **Wilcoxon signed-rank test** assumes that the variable under consideration has a symmetric distribution but does not require the distribution to be normal. The advantage to using this method is that it is resistant to outliers. However, if normality can be assumed the t-test should be used because it is more powerful than the Wilcoxon test.

Example 9.20   The U.S. Department of Agriculture estimates that a typical U.S. family of four would spend $157 per week for food. A random sample of 10 Kansas families of four yielded the weekly food costs shown in Table 9.10. At the 5% significance level, do the data provide sufficient evidence to conclude that in 1998, the mean weekly food cost for Kansas families of four was less than the national mean of $157?

Table 9.10

143	169	149	135	161
138	152	150	141	159

Solution: Let μ denote the mean weekly food cost for all Kansas families of four.

Step 1: State the null and alternative hypotheses.
H_o: μ = $157 (mean was not less than the national mean)

H_a: μ < $157 (mean was less than the national mean)

Step 2: Decide on the significance level.
We are to perform the hypothesis test at the 5% significance level; so $\alpha = 0.05$.

Step 3: Compute the value of the test statistic.
We will use the WILCOX program contained in the WeissStats CD to compute the test statistic. This program requires that the data be entered in List 1 and uses Lists 2 through 5 to run the program.

1. Press PRGM and arrow down to WILCOX and press ENTER. prgmWILCOX will appear on your home screen.

2. Press ENTER to run the program.

3. Enter the mean/median for your problem, here 157, and press ENTER.

4. The program will compute the test statistic, and the sample size and display them for you in that order. See Figure 9.13. Note the program will throw out any values that equal the mean and will display the adjusted sample size.

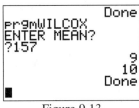

Figure 9.13

Here our test statistic is W = 9 with a sample size of 10.

Step 4: Determine the critical value.
Determine the critical value using Table VI from the textbook. Here our critical value is W = 11 because our sample size is 10. Thus we will reject if our W ≤ 11 and do not reject if W>11.

Step 5: If the test statistic falls in the rejection region, reject H_0; otherwise do not reject H_0.
Since 9 is less than 11, our critical value, we will reject H_0.

Step 6: Interpret the results of the hypothesis test.
At the 5% significance level, the data provide sufficient evidence to conclude that the mean weekly food cost for Kansas families of four was less than the national mean of $157.

9.6 Practice Problems

Problem 9.137   In 2007, the median age of U.S. residents was 36.6 years, as reported by the Census Bureau in *Current Population Reports*. A random sample taken this year of 10 U.S. residents yielded the ages, in years, shown in Table 9.11. At the 1% significance level, do the data provide sufficient evidence to conclude that the median age of today's U.S. residents has increased over the 2007 median age of 36.6 years?

Table 9.11	43	63	15	58	37
	46	50	40	12	27

Problem 9.139   The *Kelley Blue Book* provides data on retail and trade-in values for used cars and trucks. The retail value represents the price a dealer might charge after preparing the vehicle for sale. A 2006 Ford Mustang coupe has a 2009 *Kelley Blue Book* retail value of $13,015. We obtained the asking prices, in dollars, in Table 9.12 for a sample of 2006 Ford Mustang coupes for sale in Phoenix, Arizona. At the 10% significance level, do the data provide sufficient evidence to conclude that the mean asking price for 2006 Ford Mustang coupes in Phoenix is less than the 2009 *Kelley Blue Book* value?

Table 9.12	13,645	13,157	13,153	12,965	12,764
	12,664	11,665	10,565	12,665	12,765

Problem 9.140   The National Center for Health Statistics reports in *Vital Statistics of the United States* that the median birth weight of U.S. babies was *7.4* lb in 2002. A random sample of this year's births provided the weights, in pounds, given in Table 9.13. Can we conclude that this year's median birth weight differs from that in 2002? Use a significance level of *0.05*.

Table 9.13	8.6	7.4	5.3	13.8	7.8	5.7	9.2
	8.8	8.2	9.2	5.6	6.0	11.6	7.2

Problem 9.141   Two researchers, Mohammed Ajmal and Ahsan Ullah Khan, reported their findings on experiments with brewery wastes used for agricultural purposes in the article "Effects of Brewery Effluent on Agricultural Soil and Crop Plants" (*Environmental Pollution (Series A)*, 33, pp. 3441-351). Based on the results of the study, the data in Table 9.14 provides the percentages of limestone in the soil obtained by using 100% effluent. Can you conclude that the mean available limestone in soil treated with 100% MMBL effluent exceeds 2.30%, the percentage ordinarily found? Perform the test at the 1% significance level.

Table 9.14	2.41	2.31	2.54	2.28	2.72
	2.60	2.51	2.51	2.42	2.70

Problem 9.143 A manufacturer of liquid soap produces a bottle with an advertised content of 310 ml. Sixteen bottles are randomly selected and found to have the contents, in milliliters listed in Table 9.15. A normal probability plot indicates that it is reasonable to assume the contents are normally distributed. At the 5% significance level, do the data provide sufficient evidence to conclude the mean content is less than advertised?

Table 9.15	297	318	306	300
	311	303	291	298
	322	307	312	300
	315	296	309	311

LESSON 9.7 TYPE II ERROR PROBABILITIES; POWER*

As you learned in Section 9.1, hypothesis tests do not always yield correct conclusions; they have built-in margins of error. An important part of planning a study is to consider both types of errors that can be made and their effects.

Recall that two types of errors are possible with hypothesis tests. One is a Type I error: rejecting a true null hypothesis. The other is a Type II error: not rejecting a false null hypothesis. Also recall that the probability of making a Type I error is called the significance level of the hypothesis test and is denoted α, and that the probability of making a Type II error is denoted β.

In modern statistical practice, analysts generally use the probability of not making a Type II error, called the **power,** to appraise the performance of a hypothesis test. Once we know the Type II error probability, β, obtaining the power is simple—we just subtract β from 1.

The **power** of a hypothesis test is the probability of not making a Type II error, that is, the probability of rejecting a false null hypothesis. We have

$$\text{Power} = 1 - P(\text{Type II error}) = 1 - \beta.$$

Ideally, both Type I and Type II errors should have small probabilities. In terms of significance level and power, then, we want to specify a small significance level (close to 0) and yet have large power (close to 1). The smaller we specify the significance level, the smaller will be the power. However, by using a large sample, we can have both a small significance level and large power. In practice, larger sample sizes tend to increase the cost of a study. Consequently, we must balance, among other things, the cost of a large sample against the cost of possible errors.

LESSON 9.8 WHICH PROCEDURE SHOULD BE USED?*

In selecting the correct procedure, keep in mind that the best choice, is the procedure expressly designed for the distribution type under consideration, if such a procedure exists. If the sample size is large then it is appropriate to use the t-test, provided there are no outliers. If the sample size is small and the population is normal, the t-test should be used. If the sample size is small and the population is nonnormal then you should apply the Wilcoxon signed-rank test, provided the population is reasonably symmetric.

In practice, we need to look at the normal probability plot and the stem-and-leaf diagram or histogram of the sample data to ascertain the distribution type before we can select the appropriate procedure. For moderate sample sizes, histograms are preferable to stem-and-leaf diagrams for judging symmetry.

Example 9.25 According to *Food Consumption, Prices, and Expenditures*, published by the U.S. Department of Agriculture, the 2006 mean for chicken consumption was 61.3 lb per person. A simple random sample of 17 people had chicken consumption for last year as shown in Table 9.16.

Table 9.16	57	69	63	49	63	61
	72	65	91	59	0	82
	60	75	55	80	73	

Do the data provide sufficient evidence to conclude that the mean chicken consumption for last year has changed from the 2006 mean of 61.3 lbs? Which procedure should be used to perform the hypothesis test?

Solution: Use a normal probability plot, a histogram and a boxplot to determine which procedure should be used to perform a hypothesis test on the population mean.

The **normal probability plot** (Figure 9.14) shows some curvature and/or the presence of outliers. This leads to the conclusion the data is nonnormal.

Next, we need to decide if the distribution is symmetric. The sample size is moderately large so we will inspect the histogram. The histogram (Figure 9.15) shows that the data have a reasonably symmetric distribution with some possible outliers.

The boxplot (Figure 9.16) also shows the distribution to be symmetric with one outlier present. Therefore, we are led to the Wilcoxon signed-rank procedure to test the hypothesis.

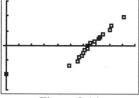

Figure 9.14

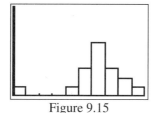

Figure 9.15

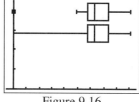

Figure 9.16

CHAPTER 10
INFERENCES FOR TWO POPULATION MEANS

LESSON 10.1   SAMPLING DISTRIBUTION OF THE DIFFERENCE BETWEEN TWO SAMPLE MEANS FOR INDEPENDENT SAMPLES

With independent simple random samples, each possible pair of samples (one from each population) is equally likely to be the pair of samples selected.

KEY FACT 10.1: Suppose that x is a normally distributed variable on each of two populations. Then, for independent samples of size n_1 and n_2 from the two populations,

$$\overline{x}_1 - \overline{x}_2 \text{ is normally distributed with mean } \mu_{\overline{x}_1 - \overline{x}_2} = \mu_{\overline{x}_1} - \mu_{\overline{x}_2}$$

$$\text{and standard deviation } \sigma_{\overline{x}_1 - \overline{x}_2} = \sqrt{\sigma_1^2/n_1 + \sigma_2^2/n_2} \ .$$

Depending on the assumptions there are three different hypothesis tests for two population means that can be used. The first test is used when both population standard deviations are known.

The other two tests are used when the two population standard deviations are unknown. In this case it is standard practice when the sample sizes are large to substitute sample estimates for the population standard deviations and treat the results as approximate.

LESSON 10.2   INFERENCES FOR TWO POPULATION MEANS, USING INDEPENDENT SAMPLES: POPULATION STANDARD DEVIATIONS ASSUMED EQUAL

Procedure 10.1 of Section 10.2 of the textbook outlines the steps for performing a pooled t-test. Procedure 10.2 outlines the steps for performing a pooled t-interval.

The assumptions are

1. Simple random samples

2. Independent samples

3. Normal populations or large samples

4. Equal population standard deviations

To test for normal populations, we can use normal probability plots. To check the equal population standard deviation assumption, we can check it informally by comparing the standard deviations of the two samples. Stem-and-leaf diagrams, histograms, or boxplots can also be viewed provided the same scale is used for both samples.

Pooled t-Test for Two Population Means

Example 10.3   The American Association of University Professors (AAUP) conducts salary studies of college professors and publishes its findings in *AAUP Annual Report on the Economic Status of the Profession*. We want to decide whether the mean salaries of college faculty teaching in public and private institutions are different. We randomly and independently sample 30 faculty members from public institutions and 35 faculty members from private institutions and these salaries are repeated in Table 10.1, where they are given in thousands of dollars rounded to the nearest hundred. At the 5% significance level, do the data provide sufficient evidence to conclude that the mean salaries for faculty teaching in public and private institutions differ?

Table 10.1	Sample 1 (public institutions)					Sample 2 (Private institutions)					
	49.9	105.7	116.1	40.3	123.1	87.3	75.9	108.8	83.9	56.6	99.2
	72.5	57.1	50.7	69.9	40.1	73.1	90.6	89.3	84.9	84.4	129.3
	73.9	92.5	99.9	95.1	57.9	148.1	132.4	75.0	98.2	106.3	131.5
	44.9	31.5	49.5	55.9	66.9	115.6	60.6	64.6	59.9	105.4	74.6
	75.9	103.9	60.3	80.1	89.7	87.2	45.1	116.6	106.7	66.0	99.6
	79.3	71.7	97.5	56.9	86.7	54.9	98.8	41.4	82.0	53.0	

Solution: First we check the conditions required for using the pooled t-test. The samples are independent and the sample sizes are large. Graphical analyses show that no49 outliers are present. Therefore Assumptions 1, 2 and 3 are met. The sample standard deviations are 23.95 and 22.21 so we can consider assumption 4 met.

Let μ_1 denote the mean salary of college faculty teaching in public institutions and μ_2 denote the mean salary of college faculty teaching in private institutions

Step 1: State the null and alternative hypotheses.

$$H_0: \ \mu_1 = \mu_2 \quad \text{(mean salaries are the same)}$$

$$H_a: \ \mu_1 \neq \mu_2 \quad \text{(mean salaries are different)}$$

Note that the hypothesis test is two-tailed.

Step 2: Decide on the significance level.
We are to perform the hypothesis test at the 5% significance level; so $\alpha = 0.05$.

Step 3: Compute the value of the test statistic.
 On the TI-83/84 Plus there are two ways to perform a pooled t-test for two population means. The first is useful if you have the data values but have not computed any of the summary statistics. This method follows.

1. With sample 1 in List 1 and sample 2 in List 2, press [STAT] and arrow over to the TEST menu.

2. Number 4 is the two-sample t-test so press [4] or arrow down to 4 and press [ENTER].

3. Highlight Data and press [ENTER].

4. Enter L_1 as List 1 by pressing [2nd] L1 [ENTER].

5. Enter L_2 as List 2 by pressing [2nd] L2 [ENTER].

6. Enter both frequencies as 1. (Recall the TI-83/84 Plus goes into Alpha mode for Frequencies so press [ALPHA] [1].)

7. Highlight the appropriate hypothesis and press [ENTER]. We will highlight $\neq \mu_2$ for our example.

8. Highlight Yes and press [ENTER] to select a pooled test. Your screen should appear as in Figure 10.1.

The TI-83/84 Plus has two displays for the answer. We will demonstrate both.

9. Highlight Calculate and press [ENTER]. Your screen will appear as in Figure 10.2. If you highlight Draw and press [ENTER], your screen will appear as in Figure 10.3.

Figure 10.1

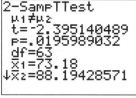
Figure 10.2

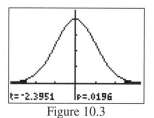
Figure 10.3

The second method of computing the test statistic and the p-value is useful if you do not have the data values but have the sample means, sample standard deviations, and sample sizes.

1. Press [STAT] and arrow over to the TEST menu.

2. Number 4 is the two-sample t-test so press [4] or arrow down to 4 and press [ENTER].

3. Highlight Stats and press [ENTER].

4. Enter the sample mean for sample 1, here 73.18, and press [ENTER].

5. Enter the sample standard deviation for sample 1, here 23.952, and press [ENTER].

6. Enter the sample size for sample 1, here 30 and press [ENTER].

7. Enter the sample mean for sample 2, here 88.19 and press [ENTER].

8. Enter the sample standard deviation for sample 2, here 26.208 and press [ENTER].

9. Enter the sample size for sample 2, here 35 and press [ENTER].

10. Highlight the appropriate hypothesis and press [ENTER]. We will highlight $\neq \mu_2$ for our example.

11. Highlight Yes and press [ENTER] to select a pooled test. Your screen should appear as in Figure 10.4

The TI-83/84 Plus has two displays for the answer. We will demonstrate both.

12. Highlight Calculate and press [ENTER]. Your screen will appear as in Figure 10.5. If you highlight Draw and press [ENTER], your screen will appear as in Figure 10.6.

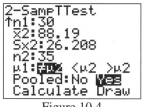
Figure 10.4

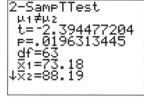
Figure 10.5

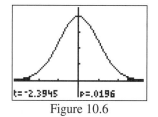
Figure 10.6

Note that both answer screens for both methods display the test statistic and the p-value. Here the test statistic is given as t = -2.394 when rounded to three decimal places and the p-value is 0.0196 when rounded to four decimal places.

When using the two different methods, your answers will vary slightly and the values may differ slightly from the answers in the textbook because the TI-83/84 Plus does not round the means and standard deviations when computing the test statistic and it does not round the test statistic to compute the p-value. However, your final conclusion will always be the same.

Step 5: If p-value $\leq \alpha$ reject H_0; otherwise do not reject H_0.
Our p-value is 0.0196 which is less than the specified significance level of 0.05. Therefore we reject the null hypothesis.

Step 6: Interpret the results of the hypothesis test.
At the 5% significance level, we have sufficient evidence to conclude that mean salaries for faculty in public and private institutions differ.

10.2 Practice Problems

Problem 10.39   The U. S. Bureau of Prisons publishes data in *Statistical Report* on the times served by prisoners released from federal institutions for the first time. Independent random samples of released prisoners in the fraud and firearms offense categories yielded the information in Table 10.2 on time served, in months. At the 5% significance level, do the data provide sufficient evidence to conclude that the mean time served for fraud is less than that for firearms offenses? Preliminary data analysis indicates that it is reasonable to assume that the data is normally distributed and that the standard deviations are approximately equal.

Table 10.2

Fraud		Firearms	
3.6	17.9	25.5	23.8
5.3	5.9	10.4	17.9
10.7	7.0	18.4	21.9
8.5	13.9	19.6	13.3
11.8	16.6	20.9	16.1

Problem 10.41   V. Tangpricha et al. conducted a study to determine whether fortifying orange juice with Vitamin D would result in changes in the blood levels of five biochemical variables. One of those variables was the concentration of parathyroid hormone (PTH), measured in picograms/milliliter (pg/mL). The researchers reported their findings in the paper "Fortification of Orange Juice with Vitamin D: A Novel Approach for Enhancing Vitamin D Nutritional Health" *American Journal of Clinical Nutrition*, Vol. 77, pp. 1478-1483). A double-blind experiment was used in which 14 subjects drank 240 mL per day of orange juice fortified with 1000 IU of Vitamin D and 12 subjects drank 240 mL per day of unfortified orange juice. Concentration levels were recorded at the beginning of the experiment and again at the end of 12 weeks. The data in Table 10.3, based on the results of the study, provide the decrease (negative values indicate increase) in PTH levels for those drinking the fortified juice and for those drinking the unfortified juice. At the 5% significance level, do the data provide sufficient evidence to conclude that drinking fortified orange juice reduces PTH level more than drinking unfortified orange juice? Preliminary data analysis indicates that it is reasonable to assume that the data is normally distributed and that the standard deviations are approximately equal.

Table 10.3

Fortified				Unfortified		
-7.7	11.2	65.8	-45.6	65.1	0.0	40.0
-4.8	26.4	55.9	-15.5	-48.8	15.0	8.8
34.4	-5.0	-2.2		13.5	-6.1	29.4
-20.1	-40.2	73.5		-20.5	-48.4	-28.7

Problem 10.51   Philosophical and health issues are prompting an increasing number of Taiwanese to switch to a vegetarian lifestyle. A study by Lu et al., published in the *Journal of Nutrition*, (Vol. 130, pp. 1591-1596) compared the daily intake of nutrients by vegetarians and omnivores living in Taiwan. Among the nutrients considered was protein. Too little protein stunts growth and interferes with all bodily functions; too much protein puts a strain on the kidneys, can cause diarrhea and dehydration, and can leach the calcium from bones and teeth. Independent samples of 51 female vegetarians and 53 female omnivores yielded the following summary statistics. Vegetarians-sample mean 39.04, sample standard deviation 18.82. Omnivores sample mean 49.92, sample standard deviation 18.97. Do the data provide sufficient evidence to conclude that the mean daily protein intakes of female vegetarians and female omnivores differ? Perform the required hypothesis test at the 1% significance level. Preliminary data analysis indicates that it is reasonable to assume that the data is normally distributed and that the standard deviations are approximately equal.

Review
Problem #7   In the paper, "Sex Differences in Static Strength and Fatigability in Three Different Muscle Groups" (*Research Quarterly for Exercise and Sport*, Vol 61(3), pp. 238-242), J. Misner et al. published results of a study on grip and leg strength of males and females. The data in Table 10.4, in newtons, is based on their measurements of right-leg strength. At the 5% significance level, do the data provide sufficient evidence to conclude that mean right-leg strength of males exceeds that of females? Preliminary data analysis indicates that it is reasonable to assume that the data is normally distributed and that the standard deviations are approximately equal.

Table 10.4

	Male			Female	
2632	1796	2256	1344	1351	1369
2235	2298	1917	2479	1573	1665
1105	1926	2644	1791	1866	1544
1569	3129	2167	2359	1694	2799
1977			1868	2098	

Pooled t-Interval Procedure for Two Population Means

Example 10.4   Consider the data in Table 10.1. Determine a 95% confidence interval for the difference, $\mu_1 - \mu_2$, between the mean salaries of faculty teaching in public and private institutions. Interpret your result.

Solution: First we check the conditions required for using the pooled t-test. The samples are independent and the sample sizes are large. Graphical analysis shows that no outliers are present. Assumptions 1, 2 and 3 are met. The sample standard deviations are 23.95 and 22.26 so we can consider assumption 4 met.

Let μ_1 denote the mean salary of college faculty teaching in public institutions and μ_2 denote the mean salary of college faculty teaching in private institutions. The TI-83/84 Plus has two different methods for computing a pooled t-interval. The first is useful if you have the data values but have not computed any of the summary statistics. This method follows.

1. With sample 1 in List 1 and sample 2 in List 2, press [STAT] and arrow over to the TEST menu.

2. Number 0 is the two-sample t-interval so press [0] or arrow down to 0 and press [ENTER].

3. Highlight Data and press [ENTER].

4. Enter L_1 as List 1 by pressing [2nd] L1 [ENTER].

5. Enter L_2 as List 2 by pressing [2nd] L2 [ENTER].

6. Enter Freq1 and Freq2 as 1. Recall the TI-83/84 Plus goes into Alpha mode for Frequencies so press ALPHA 1.

7. Enter the C-level as 0.95 and press ENTER.

8. Highlight Yes and press ENTER to select a pooled interval. Your screen should appear as in Figure 10.7.

9. Highlight Calculate and press ENTER. Your screen should appear as in Figure 10.8.

Figure 10.7

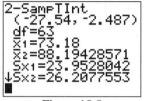
Figure 10.8

The second method of computing the interval is useful if you do not have the data values but have the sample mean, sample standard deviation and sample size.

1. With sample 1 in List 1 and sample 2 in List 2, press STAT and arrow over to the TEST menu.

2. Number 0 is the two-sample t-interval so press 0 or arrow down to 0 and press ENTER.

3. Highlight Stats and press ENTER.

4. Enter the sample mean for sample 1, here 64.48, and press ENTER.

5. Enter the sample standard deviation for sample 1, here 23.95, and press ENTER.

6. Enter the sample size for sample 1, here 30, and press ENTER.

7. Enter the sample mean for sample 2, here 73.39, and press ENTER.

8. Enter the sample standard deviation for sample 2, here 22.26, and press ENTER.

9. Enter the sample size for sample 2, here 35, and press ENTER.

10. Enter the C-level as 0.95 and press ENTER.

11. Highlight Yes and press ENTER to select a pooled interval. Your screen should appear as in Figure 10.9.

12. Highlight Calculate and press ENTER. Your screen should appear as in Figure 10.10.

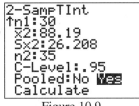
Figure 10.9

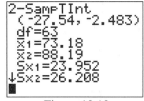
Figure 10.10

Once again, due to rounding of the t-value, sample means and sample standard deviations, the answers obtained using the two methods may differ slightly. The answers may also differ slightly from the answers in the textbook.

Here the 95% confidence interval is from -27.54 to -2.48 when rounded to two decimal places. We can be 95% confident that the difference, $\mu_1 - \mu_2$, between the mean salaries of faculty teaching in public and private institutions is somewhere between -$27,650 and -$2,480. In other words, we are 95% confident the mean salary of faculty teaching in public institutions is between $2,480 and $27,650 less than the mean salary of faculty teaching in private institutions.

10.2 Practice Problems (continued)

Problem 10.45   Consider the data in Problem 10.39. Determine a 90% confidence interval for the difference, $\mu_1 - \mu_2$, between the mean time served for fraud and firearm offenses. Interpret your result.

Problem 10.47   Consider the data in Problem 10.41. Determine a 90% confidence interval for the difference, $\mu_1 - \mu_2$, between the PTH levels for those drinking the fortified juice and for those drinking the unfortified juice.

Problem 10.51 (continued) Consider the data in Problem 10.51. Determine a 95% confidence interval for the difference, $\mu_1 - \mu_2$, between the mean daily protein intakes of female vegetarians and female omnivores. Interpret your result.

Review   Consider the data in Review Problem #7. Determine a 90% confidence interval for the difference,
Problem #8 $\mu_1 - \mu_2$, between the mean right-leg strength of males and females. Interpret your result.

LESSON 10.3 INFERENCES FOR TWO POPULATION MEANS, USING INDEPENDENT SAMPLES: POPULATION STANDARD DEVIATIONS NOT ASSUMED EQUAL

The pooled t-test and pooled t-interval require the populations' standard deviations to be equal. There are times when the population standard deviations are not equal or we do not know if they are equal or not. When this is the case, we use the **nonpooled** procedures.

Procedures 10.3 and 10.4 of Section 10.3 in the textbook outline the steps for the non-pooled t-test and non-pooled t-interval respectively. The assumptions are

1. Simple random samples

2. Independent samples

3. Normal populations or large samples

We can test for normal populations by using normal probability plots.

The Nonpooled t-Test for Two Population Means

Example 10.6   A group of neurosurgeons wanted to see whether a dynamic system (Z-plate) reduced the operative time relative to a static system (ALPS plate). R. Jacobowitz, Ph.D., an ASU professor, along with G. Vishteh, M.D., and other neurosurgeons, obtained the data in Table 10.5 on operative times, in minutes, for the two systems. At the 1% significance level, do the data provide sufficient evidence to conclude that the mean operative time is less with the dynamic system than with the static system?

Table 10.5

Dynamic							Static		
370	360	510	445	295	315	490	430	445	455
345	450	505	335	280	325	500	455	490	535

Solution: First we must check that the assumptions are satisfied. The samples are independent so Assumption 2 is satisfied. Plots reveal that there are no outliers and because the nonpooled t-test is robust to deviations from normality, we can consider Assumption 3 satisfied.

Let μ_1 denote the mean operative time for the dynamic system and μ_2 denote the mean operative time of the static system. The null and alternative hypotheses are

Step 1: State the null and alternative hypotheses.

H_0: $\mu_1 = \mu_2$ (mean dynamic time is not less than mean static time)

H_a: $\mu_1 < \mu_2$ (mean dynamic time is less than mean static time)

Step 2: Decide on the significance level.

We are to perform the hypothesis test at the 5% significance level; so $\alpha = 0.05$.

Step 3: Compute the value of the test statistic.

Note that the hypothesis test is left-tailed. On the TI-83/84 Plus there are two ways to perform a nonpooled t-test for two population means. The first is useful if you have the data values but have not computed any of the summary statistics. This method follows.

1. With sample 1 in List 1 and sample 2 in List 2, press [STAT] and arrow over to the TEST menu.

2. Number 4 is the two-sample t-test so press [4] or arrow down to 4 and press [ENTER].

3. Highlight Data and press [ENTER].

4. Enter L_1 as List 1 by pressing [2nd] L1 [ENTER].

5. Enter L_2 as List 2 by pressing [2nd] L2 [ENTER].

6. Enter both frequencies as 1. (Recall the TI-83/84 Plus goes into Alpha mode for Frequencies so press [ALPHA] [1].)

7. Highlight the appropriate hypothesis and press [ENTER]. We will highlight $< \mu_2$ for our example.

8. Highlight No and press [ENTER] to select a nonpooled test. Your screen should appear as in Figure 10.11

The TI-83/84 Plus has two displays for the answer. We will demonstrate both.

9. Highlight Calculate and press [ENTER]. Your screen will appear as in Figure 10.12. If you highlight Draw and press [ENTER], your screen will appear as in Figure 10.13.

Figure 10.11

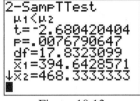
Figure 10.12

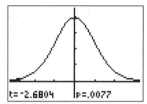
Figure 10.13

The second method of computing the test statistic and the p-value is useful if you do not have the data values but have the sample mean, sample standard deviation, and sample size.

1. With sample 1 in List 1 and sample 2 in List 2, press [STAT] and arrow over to the TEST menu.

2. Number 4 is the two-sample t-test so press [4] or arrow down to 4 and press [ENTER].

3. Highlight Stats and press [ENTER].

4. Enter the sample mean for sample 1, here 394.6, and press [ENTER].

5. Enter the sample standard deviation for sample 1, here 84.7, and press [ENTER].

6. Enter the sample size for sample 1, here 14, and press [ENTER].

7. Enter the sample mean for sample 2, here 468.3, and press [ENTER].

8. Enter the sample standard deviation for sample 2, here 38.2, and press [ENTER].

9. Enter the sample size for sample 2, here 6, and press [ENTER].

10. Highlight the appropriate hypothesis and press [ENTER]. We will highlight $< \mu_2$ for our example.

11. Highlight NO and press [ENTER] to select a nonpooled test. Your screen should appear as in Figure 10.14

The TI-83/84 Plus has two displays for the answer. We will demonstrate both.

12. Highlight Calculate and press [ENTER]. Your screen will appear as in Figure 10.15. If you highlight Draw and press [ENTER], your screen will appear as in Figure 10.16.

Figure 10.14

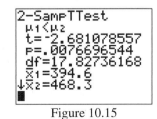
Figure 10.15

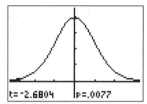
Figure 10.16

Note that both answer screens for both methods display the test statistic and the p-value. Here the test statistic is given as -2.681 when rounded to three decimal places and the p-value is 0.0077 when rounded to four decimal places. When using the two different methods, your answers will vary slightly and the values may differ slightly from the answers in the textbook because the TI-83/84 Plus does not round the means and standard deviations when computing the test statistic and it does not round the test statistic to compute the p-value. However, your final conclusion will always be the same.

Step 4: Obtain the *p*-value.
The p-value is 0.0077 when rounded to four decimal places.

Step 5: If *p*-value $\leq \alpha$ reject H_0; otherwise do not reject H_0.
Here our *p*-value of 0.0077 is less than the specified significance level of 0.01, so we can reject the null hypothesis.

Step 6: Interpret the results of the hypothesis test.
At the 1% significance level, the data provides sufficient evidence to conclude that the mean operative time for the dynamic system is less than that for the static system.

10.3 Practice Problems

Problem 10.69   According to the American Psychiatric Association, posttraumatic stress disorder (PTSD) is a common psychological consequence of traumatic events that involve threat to life or physical integrity. During the Cold War, roughly 200,000 people in East Germany were imprisoned for political reasons. Many of these prisoners were subjected to physical and psychological torture during their imprisonment, resulting in PTSD. Ehlers, Maercker, and Boos studied various characteristics of political prisoners from the former East Germany and presented their findings in the paper "Posttraumatic Stress Disorder (PTSD) Following Political Imprisonment: The Role of Mental Defeat, Alienation and Perceived Permanent Change" (*Journal of Abnormal Psychology*, Vol. 109, pp. 45-55). The researchers randomly and independently selected 32 former prisoners diagnosed with chronic PTSD and 20 former prisoners that were diagnosed with PTSD after release from prison but had since recovered (remitted). The ages, in years, at arrest yielded the data in Table 10.6. At the 10% significance level, do the data provide sufficient evidence to conclude that a difference exists in the mean age at arrest of East German prisoners with chronic PTSD and remitted PTSD? Preliminary data analysis indicates that it is reasonable to assume that the data is normally distributed.

Table 10.6

Chronic	Remitted
Mean = 25.8	Mean = 22.1
St. Dev. = 9.2	St. Dev. = 5.7
n = 32	n = 20

Problem 10.71 Several neurosurgeons wanted to see whether a dynamic system (Z-plate) reduced the number of acute postoperative days relative to a static system (ALPS plate). They obtained the data in Table 10.7. At the 5% significance level, do the data provide sufficient evidence to conclude that the mean number of acute postoperative days in the hospital are fewer with the dynamic system than with the static system? Preliminary data analysis indicates that it is reasonable to assume that the data is normally distributed.

Table 10.7

Dynamic							Static		
7	5	8	8	6	7	7	6	18	9
9	10	7	7	7	7	8	7	14	9

Problem 10.73 Previous research has suggested that changes in the activity of dopamine, a neurotransmitter in the brain, may be a causative factor of schizophrenia. In the paper "Schizophrenia: Dopamine b-Hydroxylase Activity and Treatment Response" (*Science*, Vol. 216, pp. 1423-1425), Sternberg et al. published the results of their study in which they examined 25 schizophrenic patients who had been classified as either psychotic or not psychotic by hospital staff. The activity of dopamine was measured in each patient by using the enzyme dopamine b-hydroxylase to assess differences in dopamine activity between the two groups. The data is given in Table 10.8, in nmol/ml-h/mg. At the 1% significance level, do the data suggest that dopamine activity is higher, on the average, in psychotic patients? Preliminary data analysis indicates that it is reasonable to assume that the data is normally distributed.

Table 10.8

Psychotic		Not psychotic		
0.0150	0.0222	0.0104	0.0230	0.0145
0.0204	0.0275	0.0200	0.0116	0.0180
0.0306	0.0270	0.0210	0.0252	0.0154
0.0320	0.0226	0.0105	0.0130	0.0170
0.0208	0.0245	0.0112	0.0200	0.0156

Review Problem #9 A study published by Blem and Blem in the *Journal of Herpetology* (Vol. 29, pp. 391-398) examined the reproductive characteristics of the eastern cottonmouth. The data in Table 10.9 are based on the results of the researcher's study, that give the number of young per litter for 24 female cottonmouths in Florida and 44 female cottonmouths in Virginia. At the 1% significance level, do the data provide sufficient evidence to conclude that, on the average, the number of young per litter of cottonmouths in Florida is less than that in Virginia? Preliminary data analysis indicates that it is reasonable to assume that the data is normally distributed.

Table 10.9

Florida			Virginia					
8	6	7	5	12	7	7	6	8
7	4	3	12	9	7	4	9	6
1	7	5	12	7	5	6	10	3
6	6	5	10	8	8	12	5	6
6	8	5	10	11	3	8	4	5
5	7	4	7	6	11	7	6	8
6	6	5	8	14	8	7	11	7
5	5	4	5	4				

The Nonpooled t-interval for Two Population Means

Example 10.7 Use the sample data in Example 10.6 to obtain a 98% confidence interval for the difference, $\mu_1 - \mu_2$, between the mean operative times of the dynamic and static systems. Interpret your result.

Solution First we check the conditions required for using the nonpooled t-interval. The samples are independent, so Assumption 2 is met. Graphical analysis shows that no outliers are present. Assumption 3 is met.

Let μ_1 denote the mean salary of sample 1 and μ_2 denote the mean salary of sample 2. The TI-83/84 Plus has two different methods for computing a nonpooled t-interval. The first is useful if you have the data values but have not computed any of the summary statistics. This method follows.

1. With sample 1 in List 1 and sample 2 in List 2, press $\boxed{\text{STAT}}$ and arrow over to the TEST menu.

2. Number 0 is the two-sample t-interval so press $\boxed{0}$ or arrow down to 0 and press $\boxed{\text{ENTER}}$.

3. Highlight Data and press ENTER.

4. Enter L_1 as List 1 by pressing 2nd L1 ENTER.

5. Enter L_2 as List 2 by pressing 2nd L2 ENTER.

6. Enter Freq1 and Freq2 as 1. Recall the TI-83/84 Plus goes into Alpha mode for Frequencies so press ALPHA 1.

7. Enter the C-level as 0.98 and press ENTER.

8. Highlight No and press ENTER to select a nonpooled interval. Your screen should appear as in Figure 10.17.

9. Highlight Calculate and press ENTER. Your screen should appear as in Figure 10.18.

Figure 10.17

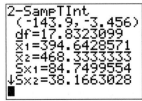
Figure 10.18

The second method of computing the interval is useful if you do not have the data values but have the sample means, sample standard deviations and sample sizes.

1. With sample 1 in List 1 and sample 2 in List 2, press STAT and arrow over to the TEST menu.

2. Number 0 is the two-sample t-interval so press 0 or arrow down to 0 and press ENTER.

3. Highlight Stats and press ENTER.

4. Enter the sample mean for sample 1, here 394.6, and press ENTER.

5. Enter the sample standard deviation for sample 1, here 84.7, and press ENTER.

6. Enter the sample size for sample 1, here 14, and press ENTER.

7. Enter the sample mean for sample 2, here 468.3, and press ENTER.

8. Enter the sample standard deviation for sample 2, here 38.2, and press ENTER.

9. Enter the sample size for sample 2, here 6, and press ENTER.

10. Enter the confidence level as 0.95 and press ENTER.

11. Highlight No and press ENTER to select a nonpooled interval. Your screen should appear as in Figure 10.19.

12. Highlight Calculate and press ENTER. Your screen should appear as in Figure 10.20.

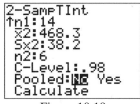

Figure 10.19

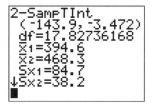
Figure 10.20

Once again, due to rounding of the t-value, sample means and sample standard deviations, the answers obtained using the two methods may differ slightly. The answers may also differ slightly from the answers in the textbook.

Here the 98% confidence interval is from -143.9 to -3.47 when rounded to two decimal places. We can be 98% confident that the difference between the mean operative times of the dynamic and static systems is somewhere between -143.9 and -3.47. In other words, we can be 98% confident that compared to the static system, the dynamic system reduces the mean operative time by somewhere between 3.47 and 143.9 minutes.

10.3 Practice Problems (continued)

Problem 10.75   Consider the data in Problem 10.69. Determine a 90% confidence interval for the difference, $\mu_1 - \mu_2$, between the mean age at arrest of East German prisoners with chronic PTSD and remitted PTSD. Interpret your result.

Problem 10.77   Consider the data in Problem 10.71. Determine a 90% confidence interval for the difference, $\mu_1 - \mu_2$, between the mean number of acute postoperative days in the hospital with the dynamic system and the static system. Interpret your result.

Problem 10.79   Consider the data in Problem 10.73. Determine a 98% confidence interval for the difference, $\mu_1 - \mu_2$, between the dopamine activity in psychotic and not psychotic patients. Interpret your result.

Review
Problem # 10   Consider the data in Review Problem #9. Determine a 98% confidence interval for the difference, $\mu_1 - \mu_2$, between the mean number of young per litter of cottonmouths in Florida and Virginia. Interpret your result.

LESSON 10.4   THE MANN-WHITNEY TEST*

The pooled and nonpooled t-tests require that the distributions of the variable under consideration be normal. The Mann-Whitney test is used when the two distributions of the variable under consideration have the same shape.

Procedure 10.5 of Section 10.4 of the textbook outlines the procedure for performing the Mann-Whitney Test. The assumptions are

1. Simple random samples

2. Independent samples

3. Same-shape populations

Example 10.11   Independent samples of employees with and without computer experience were timed to see how long it would take them to comprehend a self-study manual that explained how to use a computer to track their company's products. The times, in minutes, are given in Table 10.10. At the 5% significance level, do the data provide sufficient evidence to conclude that the mean comprehension time for employees without computer experience exceeds that for employees with computer experience?

Table 10.10

Without Experience				With Experience			
139	118	164	151	142	109	130	107
182	140	134		155	88	95	104

Solution: Remember that the smaller sample must be considered as sample 1. Therefore, Let μ_1 denote the mean time for those without experience and μ_2 denote the mean time for those with experience.

Step 1: State the null and alternative hypotheses.
   H_0: $\mu_1 = \mu_2$ (mean time for inexperienced employees is not greater)
   H_a: $\mu_1 > \mu_2$ (mean time for inexperienced employees is greater)

Step 2: Decide on the significance level.
We are to perform the hypothesis test at the 5% significance level; so $\alpha = 0.05$.

Step 3: Compute the value of the test statistic.
We will use the program MANNWHIT to compute the test statistic. Enter the samples into List 1 and List 2. Note that the program will automatically use the smaller list as sample 1 when computing the test statistic.

1. Press $\boxed{\text{PRGM}}$ and arrow down to MANNWHIT and press $\boxed{\text{ENTER}}$. prgmMANNWHIT will appear on your home screen.

2. Press $\boxed{\text{ENTER}}$ to run the program. The test statistic will be displayed on the screen.

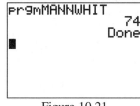
Figure 10.21

Step 4: Determine the critical value.
Determine the critical value from Table VIII of the textbook. For this example $M_r = 71$. We will reject H_0 if M exceeds 71.

Step 5: If the test statistic falls in the rejection region, reject H_0; otherwise do not reject H_0.
The value of the test statistic is 74 which exceeds the critical value of 71. Therefore, we reject H_0.

Step 6: Interpret the results of the hypothesis test.
At the 5% significance level, the data provide sufficient evidence to conclude that the mean comprehension time for employees without computer experience exceeds that for employees with computer experience.

10.4 Practice Problems

Problem 10.111   A college chemistry instructor was concerned about the detrimental effects of poor mathematics background on her students. She randomly selected 15 students and divided them according to math background. Their semester averages are listed in Table 10.11. At the 5% significance level, do the data provide sufficient evidence to conclude that, in this teacher's chemistry courses, students with fewer than two years of high-school algebra have a lower mean semester average than those with two or more years? Assume that both populations have roughly the same shape.

Table 10.11	Fewer than two years of high-school algebra		Two or more years of high-school algebra		
	58	61	84	92	75
	81	64	67	83	81
	74	73	65	52	74

Problem 10.114   The National Center for Education Statistics surveys college libraries to obtain information on the number of volumes held. Results of the surveys are published in *Digest of Education Statistics* and *Academic Libraries*. Independent random samples of public and private colleges yield the data in Table 10.12, on number of volumes held, in thousands. At the 5% significance level, do the data provide sufficient evidence to conclude that the median number of volumes held by public colleges is less than that held by private colleges? Assume that both populations have roughly the same shape.

Table 10.12	Public	79	41	516	15	24	411	265
	Private	139	603	113	27	67	500	

Problem 10.115   The U. S. Bureau of Prisons publishes data in *Statistical Report* on the times served by prisoners released from federal institutions for the first time. Independent random samples of released prisoners in the fraud and firearms offense categories yielded the information in Table 10.13 on time served, in months. At the 5% significance level, do the data provide sufficient evidence to conclude that the mean time served for fraud is less than that for firearms offenses? Assume that both populations have roughly the same shape.

Table 10.13	Fraud		Firearms	
	3.6	17.9	25.5	23.8
	5.3	5.9	10.4	17.9
	10.7	7.0	18.4	21.9
	8.5	13.9	19.6	13.3
	11.8	16.6	20.9	16.1

LESSON 10.5   INFERENCES FOR TWO POPULATION MEANS, USING PAIRED SAMPLES

Procedure 10.6 of Section 10.5 outlines the steps to perform a paired t-test. Procedure 10.7 outlines the steps to perform a paired t-interval. The assumptions are

1. Simple random paired sample

2. Normal differences or large sample

The Paired t-Test for Two Populations Means
We can use the paired t-test to test the population means. Our test statistic is

$$t = \frac{\overline{d}}{s_d / \sqrt{n}}$$

Because d-bar is the mean of the differences and the null is $\mu_1 - \mu_2$, we can use the t-test on the TI-83/84 Plus with d-bar being our x-bar, s_d our sample standard deviation, and μ_0 equal to 0.

Example 10.16   The U.S. Census Bureau publishes information on the ages of married people in *Current Population Reports*. Suppose that we want to decide whether, in the United States, the mean age of married men differs from the mean age of married women. To test that hypothesis, *10* couples are randomly selected. The resulting ages, in years, are displayed in Table 10.14. At the *5%* level of significance, do the data provide sufficient evidence to conclude that the mean age of married men differs from the mean age of married women?

Table 10.14	Husband	59	21	33	78	70	33	68	32	54	52
	Wife	53	22	36	74	64	35	67	28	41	44

Note the samples are *paired* since for each couple, the age of the husband is paired with the age of the wife. Let μ_1 denote the mean age of all married men and μ_2 denote the mean age of all married women.

Step 1: State the null and alternative hypotheses.

H_0: $\mu_1 = \mu_2$ (mean age is the same)

H_a: $\mu_1 \neq \mu_2$ (mean age differs)

Note that this hypothesis test is two tailed.

The first step is to check the assumptions and determine if they are met. To do this, we must first find the differences and then test the differences for normality.

1. Enter the first sample into List 1. (Here that is the ages of the married men.)

2. Enter the second sample into List 2 (Here that is the ages of the married women.)

3. From the home screen, find the list of differences by pressing [2nd] L1 [–] [2nd] L2 [STO▶] [2nd] L3 [ENTER].

4. Test List 3 (the differences) for normality using a normal probability plot. Figure 10.22 shows the normal probability plot of the differences are approximately linear so the assumption of normality is met. Thus, we can proceed with the hypothesis test.

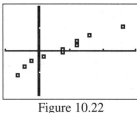
Figure 10.22

Step 2: Decide on the significance level.
We are to perform the hypothesis test at the 5% significance level; so $\alpha = 0.05$.

Step 3: Compute the value of the test statistic.
There are two ways to perform a paired t-test using the TI-83's one-sample t-test. The first is useful if you have the data values but have not computed the sample difference mean and sample difference standard deviation. We will use this method first.

5. Press $\boxed{\text{STAT}}$ and arrow over to the TEST menu.

6. Number 2 is the T-test so press $\boxed{2}$ or arrow down to 2 and press $\boxed{\text{ENTER}}$.

7. Highlight Data and press $\boxed{\text{ENTER}}$.

8. Enter μ_0 as 0 and press $\boxed{\text{ENTER}}$.

9. Enter List as List 3 by pressing $\boxed{\text{2nd}}$ L3 $\boxed{\text{ENTER}}$.

10. Enter Freq: as 1 Recall the TI-83/84 Plus goes into Alpha mode for Frequencies so press $\boxed{\text{ALPHA}}$ $\boxed{1}$.

11. Highlight the appropriate alternative hypothesis and press $\boxed{\text{ENTER}}$. We will highlight $\neq \mu_0$ for our example. Your screen should appear as in Figure 10.23.

The TI-83/84 Plus has two displays for the answer. We will demonstrate both.

12. Highlight Calculate and press $\boxed{\text{ENTER}}$. Your screen will appear as in Figure 10.24. If you highlight Draw and press $\boxed{\text{ENTER}}$, your screen will appear as in Figure 10.25.

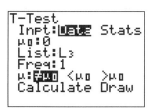

Figure 10.23

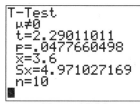

Figure 10.24

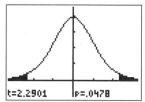

Figure 10.25

The second method of computing the test statistic and the p-value is useful if you do not have the data values but have the sample difference mean, sample difference standard deviation and sample size.

1. Press $\boxed{\text{STAT}}$ and arrow over to the TEST menu.

2. Number 2 is the T-test so press $\boxed{2}$ or arrow down to 2 and press $\boxed{\text{ENTER}}$.

3. Highlight Stats and press $\boxed{\text{ENTER}}$.

4. Enter μ_0 as 0 and press $\boxed{\text{ENTER}}$.

5. Enter the sample difference mean, here 3.6, and press $\boxed{\text{ENTER}}$.

6. Enter the sample difference standard deviation, here 4.971, and press $\boxed{\text{ENTER}}$.

7. Enter the sample size, here 10, and press $\boxed{\text{ENTER}}$.

8. Highlight the appropriate alternative hypothesis and press $\boxed{\text{ENTER}}$. We will highlight $> \mu_0$ for our example. Your screen should appear as in Figure 10.26.

The TI-83/84 Plus has two displays for the answer. We will demonstrate both.

9. Highlight Calculate and press $\boxed{\text{ENTER}}$. Your screen will appear as in Figure 10.27. If you highlight Draw and press $\boxed{\text{ENTER}}$, your screen will appear as in Figure 10.28.

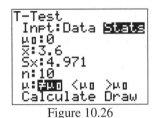

Figure 10.26

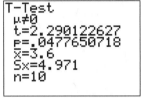

Figure 10.27

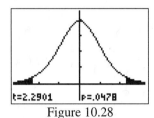

Figure 10.28

Note that both answer screens for both methods display the test statistic and the p-value. Here the test statistic is given as 4.71 when rounded to two decimal places and the p-value is .0006 when rounded to 3 decimal places. When using the two different methods, your answers will vary slightly and the values may differ slightly from the answers in the textbook because the TI-83/84 Plus does not round the mean to compute the test statistic. The TI-83/84 Plus also allows you to calculate a more exact value of p than the table does.

Step 4: Obtain the *p*-value.
The p-value is 0.0478 when rounded to four decimal places.

Step 5: If *p*-value $\leq \alpha$ reject H$_0$; otherwise do not reject H$_0$.
Since the *p*-value is less than the specified significance level of 0.05, we reject the null hypothesis.

Step 6: Interpret the results of the hypothesis test.
At the 5% significance level, the data provide sufficient evidence to conclude that the mean age of all married men differs from the mean age of all married women.

10.5 Practice Problems

Problem 10.141   Charles Darwin, author of *Origin of Species*, investigated the effect of cross-fertilization on the heights of plants. In one study he planted 15 pairs of Zea mays plants. Each pair consisted of one cross-fertilized plant grown in the same pot. Table 10.15 gives the height differences, in eighths of an inch, for the 15 pairs. Each difference is obtained by subtracting the height of the self-fertilized plant from that of the cross-fertilized plant. At the 5% significance level, do the data provide sufficient evidence to conclude that the mean heights of cross-fertilized and self-fertilized Zea mays differ? Preliminary data analyses indicate that we can use the paired t-test.

Table 10.15				
49	-67	8	16	6
23	28	41	14	29
56	24	75	60	-48

Problem 10.143 Anorexia nervosa is a serious eating disorder, found particularly among young women. The data in Table 10.16 provide the weights, in pounds, of 17 anorexic young women before and after receiving a family-therapy treatment for anorexia nervosa. [SOURCE: Hand et al. (ed.) *A Handbook of Small Data Sets*, London:Chapman & Hall, 1994. Raw data from B. Everitt]. Does it appear that family therapy is effective in helping anorexic young women gain weight, on average? Perform the appropriate hypothesis test at the 5% significance level. Preliminary data analyses indicate that we can use the paired t-test.

Table 10.16

Before	After	Before	After	Before	After
83.3	94.3	76.9	76.8	82.1	95.5
86.0	91.5	94.2	101.6	77.6	90.7
82.5	91.9	73.4	94.9	83.5	92.5
86.7	100.3	80.5	75.2	89.9	93.8
79.6	76.7	81.6	77.8	86.0	91.7
87.3	98.0	83.8	95.2		

Problem 10.145 Glaucoma is a leading cause of blindness in the United States. N. Ehlers measured the corneal thickness of eight patients who had glaucoma in one eye but not in the other. The results of the study were published as the paper "On Corneal Thickness and Intraocular Pressure, II" (*Acta Opthalmological*, Vol. 48, pp. 1107-1112). The data is given in Table 10.17 in microns. At the 10% significance level, do the data provide sufficient evidence to conclude that mean corneal thickness is greater in normal eyes than in eyes with glaucoma? Preliminary data analyses indicate that we can use the paired t-test.

Table 10.17

Patient	Normal	Glaucoma
1	484	488
2	478	478
3	492	480
4	444	426
5	436	440
6	398	410
7	464	458
8	476	460

The Paired T-interval Procedure for Two Population Means

We can use the one-sample t-interval on the TI-83/84 Plus to compute a paired t-interval. Let $\bar{d}$ be our sample mean, and s_d be our sample standard deviation.

Example 10.17 The U.S. Census Bureau publishes information on the ages of married people in *Current Population Reports*. Suppose that we want to decide whether, in the United States, the mean age of married men differs from the mean age of married women. To test that hypothesis, *10* couples are randomly selected. The resulting ages, in years, are displayed in Table 10.18. Obtain a 95% confidence interval for the difference between the mean age of married men and married women.

Table 10.18

Husband	59	21	33	78	70	33	68	32	54	52
Wife	53	22	36	74	64	35	67	28	41	44

Note the samples are *paired* since for each couple, the age of the husband is paired with the age of the wife. Let μ_1 denote the mean age of all married men and μ_2 denote the mean age of all married women.

Solution: In Example 10.16 we verified the assumptions are met. To obtain a 95% confidence interval for the difference, $\mu_1 - \mu_2$,

1. Enter the first sample into List 1. (Here that is ages of the married men.)

2. Enter the second sample into List 2 (Here that is ages of the married women.)

There are two ways to do a t-interval on the TI-83. The first is best used when you have the data values, but do not have the sample difference mean and sample difference standard deviation.

5. With the data in List 3, press STAT and arrow over to the TESTS menu.

6. Number 8 is the TInterval so either press 8 or arrow down to 8 and press ENTER.

7. Highlight Data and press ENTER.

8. Enter List as List 3 by pressing 2nd L3 ENTER.

9. Enter the Freq: as 1. Recall the TI-83/84 Plus goes into Alpha mode for Frequencies so press ALPHA 1.

10. Enter the C-level as 0.90. Your screen should appear as in Figure 10.29.

11. Highlight Calculate and press ENTER. Your answer will be displayed as in Figure 10.30.

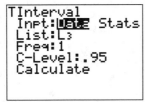
Figure 10.29

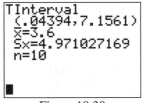
Figure 10.30

The second method of computing the interval is useful if you do not have the data values but have the sample difference mean and sample difference standard deviation.

1. Press STAT and arrow over to the TESTS menu.

2. Number 8 is the TInterval so either press 8 or arrow down to 8 and press ENTER.

3. Highlight Stats and press ENTER.

4. Enter your sample mean as 3.6.

5. Enter your sample standard deviation as 4.971.

6. Enter your sample size as 10.

7. Enter the C-level as 0.95. Your screen should appear as in Figure 10.31.

8. Highlight Calculate and press ENTER. Your answer will be displayed as in Figure 10.32.

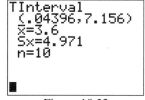

Figure 10.31	Figure 10.32

Once again, due to rounding of the t-value, sample means and sample standard deviations, the answers obtained using the two methods may differ slightly. The answers may also differ slightly from the answers in the textbook.

The 90% confidence interval is seen to be (0.04394, 7.1560), that is, we are 90% confident that the mean age of married men is between 0.04 and 7.16 years greater than the mean age of married women.

10.5 Practice Problems (continued)

Problem 10.147 Consider the data in Problem 10.141. Determine a 95% confidence interval for the difference, $\mu_1 - \mu_2$, between the mean heights of cross-fertilized and self-fertilized Zea mays. Interpret your result.

Problem 10.149 Consider the data in Problem 10.143. Determine a 90% confidence interval for the difference, $\mu_1 - \mu_2$, between the mean weights of anorexic young women before and after receiving family-therapy treatment. Interpret your result.

Problem 10.151 Consider the data in Problem 10.145. Determine a 80% confidence interval for the difference, $\mu_1 - \mu_2$, between the mean corneal thickness of normal eyes and eyes with glaucoma. Interpret your result.

LESSON 10.6   THE PAIRED WILCOXON SIGNED-RANK TEST*

The paired t-test requires that the paired-difference variable be approximately normally distributed or the sample size to be large. If this variable is not normally distributed but is symmetric we can use the **paired Wilcoxon signed-rank test**. Procedure 10.8 of Section 10.6 outlines the procedure for conducting a paired Wilcoxon signed-rank test. The assumptions are

1. Simple random paired sample

2. Symmetric differences

Example 10.19 The U.S. Census Bureau publishes information on the ages of married people in *Current Population Reports*. Suppose that we want to decide whether, in the United States, the mean age of married men differs from the mean age of married women. To test that hypothesis, *10* couples are randomly selected. The resulting ages, in years, are displayed in Table 10.19. At the *5%* level of significance, do the data provide sufficient evidence to conclude that the mean age of married men differs from the mean age of married women?

Table 10.19	Husband	59	21	33	78	70	33	68	32	54	52
	Wife	53	22	36	74	64	35	67	28	41	44

Note the samples are *paired* since for each couple, the age of the husband is paired with the age of the wife. Let μ_1 denote the mean age of all married men and μ_2 denote the mean age of all married women.

Step 1: State the null and alternative hypotheses.

H_0: $\mu_1 = \mu_2$ (mean age is the same)

H_a: $\mu_1 \neq \mu_2$ (mean age differs)

Note that this hypothesis test is two tailed.

Step 2: Decide on the significance level.

We are to perform the hypothesis test at the 5% significance level; so $\alpha = 0.05$.

Step 3: Compute the value of the test statistic.

We must first calculate the paired differences of the sample pairs.

1. Enter the first sample into List 1. (Here that is the ages of the married men.)

2. Enter the second sample into List 2 (Here that is the ages of the married women.)

3. From the home screen, find the list of differences by pressing [2nd] L1 [−] [2nd] L2 [STO▸] [2nd] L3 [ENTER]. The differences are now stored in List 3.

We will use the WILCOX program to compute the test statistic. This program requires that the paired differences be entered in List 1. We will consider our mean to be 0 for this test.

5. Move the paired differences from List 3 to List 1. On the home screen press [2nd] L3 [STO▸] [2nd] L1.

6. Press [PRGM] and arrow down to WILCOX and press [ENTER]. prgmWILCOX will appear on your screen.

7. Enter 0 for the mean/median and press [ENTER].

8. The program will compute the test statistic along with the current sample size and display them for you. See Figure 10.33.

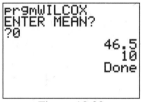

Figure 10.33

Step 4: Determine the critical values.

Now we must determine the critical values using Table VI in the textbook. The critical values for a two tailed test at the 5% significance level are $W_{0.975} = 8$ and $W_{0.025} = 47$. Thus, we will reject H_0 if the test statistic is less than or equal to 8 or if the test statistic is greater than or equal to 47.

Step 5: If the test statistic falls in the rejection region, reject H_0; otherwise do not reject H_0.

Here our test statistic is 46.5 which is in the non-rejection region. Therefore, we do not reject H_0.

Step 6: Interpret the results of the hypothesis test.

At the 5% significance level, the data does not provide sufficient evidence to conclude that the mean age of married men differs from the mean age of married women.

10.6 Practice Problems

Problem 10.173 Charles Darwin, author of *Origin of Species*, investigated the effect of cross-fertilization on the heights of plants. In one study he planted 15 pairs of Zea mays plants. Each pair consisted of one cross-fertilized plant grown in the same pot. Table 10.20 gives the height differences, in eighths of an inch, for the 15 pairs. Each difference is obtained by subtracting the height of the self-fertilized plant from that of the cross-fertilized plant. At the 1% significance level, do the data provide sufficient evidence to conclude that the mean heights of cross-fertilized and self-fertilized Zea mays differ? Preliminary data analyses indicate we can use the paired Wilcoxon Signed-Rank Test.

Table 10.20	49	-67	8	16	6
	23	28	41	14	29
	56	24	75	60	-48

Problem 10.175 Anorexia nervosa is a serious eating disorder, found particularly among young women. The data in Table 10.21 provide the weights, in pounds, of 17 anorexic young women before and after receiving a family-therapy treatment for anorexia nervosa. [SOURCE: Hand et al. (ed.) *A Handbook of Small Data Sets*, London:Chapman & Hall, 1994. Raw data from B. Everitt]. Does it appear that family therapy is effective in helping anorexic young women gain weight? Perform the appropriate hypothesis test at the 5% significance level. Preliminary data analyses indicate we can use the paired Wilcoxon Signed-Rank Test.

Table 10.21	Before	After	Before	After	Before	After
	83.3	94.3	76.9	76.8	82.1	95.5
	86.0	91.5	94.2	101.6	77.6	90.7
	82.5	91.9	73.4	94.9	83.5	92.5
	86.7	100.3	80.5	75.2	89.9	93.8
	79.6	76.7	81.6	77.8	86.0	91.7
	87.3	98.0	83.8	95.2		

Problem 10.177 Glaucoma is a leading cause of blindness in the United States. N. Ehlers measured the corneal thickness of eight patients who had glaucoma in one eye but not in the other. The results of the study were published as the paper "On Corneal Thickness and Intraocular Pressure, II" (*Acta Opthalmological*, Vol. 48, pp. 1107-1112). The data is given in Table 10.22 in microns. At the 10% significance level, do the data provide sufficient evidence to conclude that mean corneal thickness is greater in normal eyes than in eyes with glaucoma? Preliminary data analyses indicate we can use the paired Wilcoxon Signed-Rank Test.

Table 10.22	Patient	Normal	Glaucoma
	1	484	488
	2	478	478
	3	492	480
	4	444	426
	5	436	440
	6	398	410
	7	464	458
	8	476	460

Review
Problem #14   In the article "Comparison of Fiber Counting by TV Screen and Eyepieces of Phase Contrast
Microscopy" (*American Industrial Hygiene Association Journal*, Vol. 63, pp. 756-761). I. Moa, H.
Yeh, and M. Chen reported on determining fiber density by two different methods. The fiber
density of 10 samples with varying fiber density was obtained by using both an eyepiece method
and a TV-screen method. The results, in fibers per square milliliter, are presented in Table 10.23.
Use the paired Wilcoxon signed-rank test to decide whether, on average, the eyepiece method gives
a greater fiber-density reading than the TV-screen method. Perform the required hypothesis test at
the 5% significance level. Preliminary data analyses indicate we can use the paired Wilcoxon
Signed-Rank Test.

Table 10.23

Sample ID	Eyepiece	TV Screen
1	182.2	177.8
2	118.5	116.6
3	100.0	92.4
4	161.3	145.0
5	42.7	38.9
6	299.1	226.3
7	547.8	514.6
8	437.3	458.1
9	174.4	159.2
10	85.4	86.6

LESSON 10.7 WHICH PROCEDURE SHOULD BE USED?*

This chapter contains several inferential procedures for comparing the means of two populations. In selecting the
correct procedure, keep in mind that the best choice is the procedure expressly designed for the type of distribution
under consideration, if such a procedure exists, and that the three t-tests are only approximately correct for large
samples from nonnormal populations.

You should examine the sample data to settle on distribution type before choosing a procedure. SPSS can be used to
construct normal probability plots, stem-and-leaf diagrams, histograms, boxplots to examine the sample data.

CHAPTER 11
INFERENCES FOR POPULATION STANDARD DEVIATIONS*

LESSON 11.1   INFERENCES FOR ONE POPULATION STANDARD DEVIATION*

A variable is said to have a chi-square distribution if its distribution has a chi-square (χ^2) curve. The particular χ^2-curve is specified by stating its number of degrees of freedom, similar to the t-distribution. The TI-83/84 Plus has a built-in function to find the area under a χ^2-curve with specified degrees of freedom to the left of a χ^2-value.

Example: For a χ^2-curve with df = 15, find the area under the curve to the left of 6.262.

Solution

    1. Press 2nd DISTR to access the distribution menu.

    2. Press 7 or arrow down to **7: χ^2 cdf(** and press ENTER.

    3. The format of this command is χ^2 **cdf**(lowerbound, upperbound, df). For our problem we need to find the area from 0 to 6.262. For our problem our command is therefore χ^2 **cdf**(0, 6.262, 15). The result is shown in Figure 11.1.

Figure 11.1

If you are looking for the area to the right of a χ^2-value, you will need to estimate ∞ as 1E99. Keep in mind that the χ^2 curve is not symmetric, but the total area under the curve is still equal to 1.

The TI-83/84 Plus does not have a built-in Inv χ^2 function like InvNorm. However, the program INVCHI contained in the WeissStats CD can be used to accomplish the same thing.

Example 11.1   For a χ^2-curve with df = 12, find $\chi^2_{0.025}$; that is, find the χ^2-value having area 0.025 to its right.

Solution:

    1. Press PRGM and arrow down to INVCHI and press ENTER. prgmINVCHI should appear on your home screen.

    2. Press ENTER to run the program.

    3. Follow the prompts entering 12 for the DF, 2 for area to the right, and 0.025 for the area. The calculator will take a little while to do the computation. See Figure 11.2 for final result.

Figure 11.2

The χ^2-value having area 0.025 to its right is 23.337 when rounded to three decimal places.

11.1 Practice Problems

Problem 11.5   For a χ^2-curve with 19 degrees of freedom, find the χ^2-value that has area
  a) 0.025 to its right   b) 0.95 to its right.

Problem 11.7   For a χ^2-curve with df = 10, determine
  a) $\chi^2_{0.05}$ b) $\chi^2_{0.975}$

Problem 11.9   Consider the χ^2-curve with df = 8. Obtain the χ^2-value that has area
  a) 0.01 to its left   b) 0.95 to its left.

Problem 11.11 Determine the two χ^2-values that divide the area under the curve into a middle 0.95 area and two
  outside 0.025 areas for a χ^2 curve with
  a) df = 5   b) df = 26,

The χ^2-Test for a Population Standard Deviation

Procedure 11.1 of Section 11.1 outlines the steps for performing a χ^2-Test for a population standard deviation. The assumptions are:

  1. Simple random sample

  2. Normal population

The TI-83/84 Plus does not have a built-in procedure for this test. However, the p-values may be computed using the STDEVHT program contained on the WeissStats CD.

Example 11.6   Xenical is used to treat obesity in people with risk factors such as diabetes, high blood pressure, and high cholesterol or triglycerides. Xenical works in the intestines, where it blocks some of the fat a person eats from being absorbed. A standard prescription of Xenical is given in 120-milligram (mg) capsules. Although the capsule weights can vary somewhat from 120 mg and also from each other, keeping the variation small is important for various medical reasons. Based on standards set by the United States Pharmacopeia (USP)—an official public standards-setting authority for all prescription and over-the-counter medicines and other health care products manufactured or sold in the United States—we determined that a standard deviation of Xenical capsule weights of less than 2 mg is acceptable. A sample of 10 Xenical capsules had the weights, in milligrams (mg), shown in Table 11.1. At the 5% significance level, do the data provide sufficient evidence to conclude that the standard deviation of the weights of all Xenical capsules is less than 2.0 mg?

Table 11.1	120.94	118.58	119.41	120.23
	121.13	118.22	119.71	121.09
	120.56	119.11		

Solution: We begin by entering the data into a list called WEIGH.

Step 1: State the null and alternative hypotheses.
Let σ denote the standard deviation of all Xenical capsules.

$H_0 : \sigma = 2.0$ (too much weight variation)

$H_a : \sigma < 2.0$ (not too much weight variation)

Step 2: Decide on the significance level, α.
The test is to be performed at the 5% significance level. Thus $\alpha = 0.05$.

Step 3: Compute the value of the test statistic.
We will use the STDEVHT program contained in the WeissStats CD to compute the test statistics and p-value. This program allows you to use any list for the data but will place the data in List 1 to run the program.

1. Press PRGM and arrow down to STDEVHT and press ENTER. prgmSTDEVHT will appear on your home screen.

2. Press ENTER to run the program.

Similar to the built-in procedures, STDEVHT allows you to choose between entering data as a list or just using the statistics. We will demonstrate both methods. The first is useful if you have the data values but have not computed the sample standard deviation. We will use this method first.

3. Enter a 1 and press ENTER to choose the DATA option.

4. Enter your population standard deviation, here 2.0, and press ENTER.

5. Choose the appropriate alternative hypothesis by entering 1 for <, 2 for ≠ and 3 for >. We wish to conduct a left tailed test for our example, so we will enter 1 and press ENTER.

6. Enter the name of the list, WEIGH, by pressing 2nd LIST, then arrow to the name of the list and pressing ENTER.

The program will compute the test statistic and the p-value and display them for you. Here our test statistic is $\chi^2 = 2.504$.

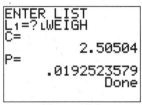

Figure 11.3

Step 4:   Obtain the p-value.

This is a left-tailed test, therefore the p-value is the area to the *left* of $\chi^2 = 2.504$ under a χ^2-curve with 9 degrees of freedom. The p-value is 0.0193 rounded to 4 decimal places.

Step 5: If p-value $\leq \alpha$, reject H$_0$; otherwise, do not reject H$_0$.

The p-value = 0.0192 is less than or equal to the significance level of 0.05, therefore we reject the null hypothesis.

Step 6: Interpret the results of the hypothesis test.

At the 5% significance level, we have sufficient evidence to conclude that the standard deviation, σ, of all Xenical capsules is less than 2 mg.

The second method of computing the test statistic and the p-value is useful if you do not have the data values but have the sample standard deviation and sample size.

1. After beginning the execution of the program, enter a 2 and press ENTER to choose the STATS option.

2. Enter your population standard deviation, here 2.0, and press ENTER.

3. Choose the appropriate alternative hypothesis by entering 1 for <, 2 for ≠ and 3 for >. We wish to conduct a left tailed test for our example, so we will enter 1 and press ENTER.

4. Enter your sample standard deviation, here 1.055, and press ENTER.

5. Enter your sample size, here 10, and press ENTER.

The program will compute the test statistic and the p-value and display them for you. See Figure 11.4.

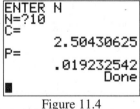

Figure 11.4

Here our test statistic is again $\chi^2 = 2.504$ and our p-value is .0192 when rounded to four decimal places.

When using the two different methods, your answers will vary slightly and the values may differ slightly from the answers in the textbook because the TI-83/84 Plus does not round anything when computing the test statistic and p-value.

11.1 Practice Problems (continued)

Problem 11.23   R. Morris and E. Watson studied various aspects of process capability in the paper "Determining Process Capability in a Chemical Batch Process" (*Quality Engineering*, Vol. 10(2), pp. 389-396). In one part of the study, the researchers compared the variability in product of a particular piece of equipment to a known analytical capability to decide whether product consistency could be improved. The data in Table 11.2 were obtained for 10 batches of product. At the 1% significance level, do the data provide sufficient evidence to conclude that the process variation for this piece of equipment exceeds the analytical capability of 0.27?

Table 11.2	30.1	30.7	30.2	29.3	31.0
	29.6	30.4	31.2	28.8	29.8

Problem 11.25   A coffee machine is supposed to dispense 6 fl oz of coffee into a paper cup. In reality, the amounts dispensed vary from cup to cup. However, if the machine is working properly, then most of the cups will contain within 10% of the advertised 6 fl oz. This means that the standard deviation of the amounts dispensed should be less than 0.2 fl oz. A random sample of 15 cups provided the data in Table 11.3. At the 5% significance level, do the data provide sufficient evidence to conclude that the standard deviation of the amounts being dispensed is less than 0.2 fl oz?

Table 11.3	5.90	5.82	6.20	6.09	5.93
	6.18	5.99	5.79	6.28	6.16
	6.00	5.85	6.13	6.09	6.18

Review Problem #9   IQs measured on the Stanford Revision of the Binet-Simon Intelligence Scale are supposed to have a standard deviation of 16 points. Twenty-five randomly selected people were given the IQ test and the data is given in Table 11.4. Preliminary data analyses and other information indicate that it is reasonable to presume that IQs measured on the Stanford Revision of the Binet-Simon Intelligence Scale are normally distributed. Do the data provide sufficient evidence to conclude that IQs measured on this scale have a standard deviation different from 16 points? Perform the test at the 10% significance level.

Table 11.4	91	96	106	116	97
	102	96	124	115	121
	95	111	105	101	86
	88	129	112	82	98
	104	118	127	66	102

The χ^2-Interval for a Population Standard Deviation

Procedure 11.2 of Section 11.1 outlines the steps for determining a χ^2-Interval for a population standard deviation. The assumptions are:

1. Simple random sample

2. Normal population

The TI-83/84 Plus does not have a built-in procedure for this interval. However, the interval may be computed using the STDEVINT program contained on the WeissStats CD.

Example 11.7   Use the sample data in Table 11.1 to determine a 95% confidence interval for the standard deviation, σ, of the weights of all Xenical capsules.

Solution: We can use the program STDEVINT to obtain the required confidence interval.

1. Press [PRGM] and arrow down to STDEVINT and press [ENTER]. prgmSTDEVINT will appear on your home screen.

2. Press [ENTER] to run the program.

3. Enter your alpha value, here 0.05, and press [ENTER].

Similar to the built-in procedures, STDEVINT allows you to choose between entering data as a list or just using the statistics. We will demonstrate both methods. The first is useful if you have the data values but have not computed the sample standard deviation. We will use this method first and assume the data is in a list named WEIGH.

4. Enter a 1 and press [ENTER] to choose the DATA option.

5. Enter the name of the list, here WEIGH by pressing [2nd] LIST, arrow to the name of the list and pressing [ENTER].

The program will compute and display the endpoints of your confidence interval. See Figure 11.5.

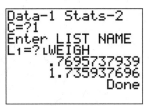

Figure 11.5

Here the endpoints are 0.770 and 1.736 when rounded to three decimal places.

The second method of computing the interval is useful if you do not have the data values but have the sample standard deviation and sample size.

1. After beginning the execution of the program and entering your alpha value, enter a 2 and press [ENTER] to choose the STATS option.

2. Enter your sample size, here 12, and press [ENTER].

3. Enter your sample standard deviation, here 1.055, and press [ENTER].

The program will compute and display the endpoints of your confidence interval. See Figure 11.6.

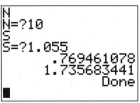

Figure 11.6

Here the endpoints are 0.769 and 1.736 when rounded to three decimal places. When using the two different methods, your answers may vary slightly and the values may vary slightly from the answers in the textbook because the TI-83/84 Plus does not round anything when computing the interval.

We can be 90% confident that the standard deviation of the weights of all Xenical capsules is somewhere between 0.769 mg and 1.736 mg.

11.1 Practice Problems (Continued)

Problem 11.29 Use the data from problem 11.23 to obtain a 98% confidence interval for the process variation of the piece of equipment under consideration.

Problem 11.31 Use the data from problem 11.25 to obtain a 90% confidence interval for the standard deviation of the amounts being dispensed in the coffee cups.

Review   Use the data in Review Problem #9 to obtain a 90% confidence interval for the standard deviation
Problem #10   of the IQs measured on the Stanford Revision of the Binet-Simon Intelligence Scale.

LESSON 11.2 INFERENCES FOR TWO POPULATION STANDARD DEVIATIONS, USING INDEPENDENT SAMPLES*

The F-Distribution

A variable is said to have an **F-distribution** if its distribution has the shape of a special type of right-skewed curve, called an **F-curve**. The particular F-curve is specified by stating its number of degrees of freedom. However, the F-distribution has two numbers of degrees of freedom instead of one.

The first number of degrees of freedom for an F-curve is called the **degrees of freedom for the numerator(dfn)** and the second the **degrees of freedom for the denominator (dfd)**. We denote the degrees of freedom for an F-curve we write df = (dfn, dfd).

The TI-83/84 Plus has a built-in function to find the area under a F-curve with specified degrees of freedom to the left of a F-value.

Example   For an F-curve with df = (3, 10) find the area to the left of 3.71.

Solution:

1. Press [2nd] DISTR to access the distribution menu.

2. Press [9] or arrow down to **9:Fcdf(** and press [ENTER].

3. The format for this command is **Fcdf**(lowerbound, upperbound, dfn, dfd). For our problem we need to find the area from 0 to 3.71. For our problem our command is therefore **Fcdf**(0, 3.71, 3, 10). The result is shown in Figure 11.7.

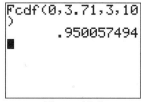

Figure 11.7

Thus, the area to the left of 3.71 for an F-curve with df = (3,10) is .950 rounded to three decimal places.

If you are looking for the area to the right of a F-value, you will need to estimate ∞ as 1E99. Keep in mind that the F-curve is not symmetric unlike the normal and t-curves, but the total area under the curve is still equal to 1. The TI-83/84 Plus does not have a built-in InvF function like InvNorm. However, the program INVF contained in the WeissStats CD can be used to accomplish the same thing.

Example 11.9 For an F-curve with df = (4, 12), find $F_{0.05}$; that is the F-value having area 0.05 to its right.

Solution:

1. Press PRGM and arrow down to INVF and press ENTER. prgmINVF will appear on your home screen.

2. Press ENTER to run the program.

3. Follow the prompts entering 4 for the df of the numerator, 12 for the df of the denominator, 2 for area to the right, and 0.05 for the area. The calculator will take a little while to do the computation. See Figure 11.8 for the final result.

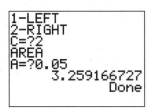

Figure 11.8

For an F-curve with df = (4, 12), the F-value having area 0.05 to the right is 3.26 when rounded to two decimal places.

11.2 Practice Problems

Problem 11.53 An F-curve has df = (24, 30), find the F-value that has the specified area to its right
   a) 0.05   b) 0.01   c) 0.025

Problem 11.55 For an F-curve with df = (20, 21), find
   a) $F_{0.01}$ b) $F_{0.05}$ c) $F_{0.10}$

Problem 11.57 For an F-curve with df = (6, 8), find the F-value that has the specified area to its left
   a) 0.01   b) 0.95

Problem 11.59 Determine the two F-values that divide the area under the curve into a middle 0.95 area and two outside 0.025 areas for an F-curve with
   a) df = (7,4)   b) df = (12,20)

Hypothesis Tests for Two Population Standard Deviations
Variation within a method used for testing a product is an essential factor in deciding whether the method should be employed. Indeed, when the variation of such a test is high, it is difficult to ascertain the true quality of a product. Therefore, it may be desirable to test whether or not two population standard deviations are equal. This test is called the **F-test for two population standard deviations**. It is also referred to as the F-test for two population variances.

Procedure 11.3 of Section 11.2 outlines the steps for performing a F-test for two population standard deviation. The assumptions are

1. Simple random samples

2. Independent samples

3. Normal populations

Unlike z-tests and t-tests for one and two population means, the F-test for two population standard deviations is not robust to moderate violations of the normality assumption. In fact, it is so nonrobust that many statisticians advise against using it unless there is considerable evidence that the variable under consideration is normally distributed on each population or very nearly so.

Consequently, before using the F-test for two population standard deviations, we should construct a normal probability plot of each sample. If either plot creates any doubt about the normality of the variable under consideration, then the F-test for two population standard deviations should not be used.

Example 11.14   In the paper "Using Repeatability and Reproducibility Studies to Evaluate a Destructive Test Method" (*Quality Engineering*, Vol. 10(2), pp. 283-290), A. Phillips et al., studied the variability of the Elmendorf tear test. That test is used to evaluate material strength for fiberglass shingles, paper quality, and other manufactured products. In one aspect of the study, the researchers independently and randomly obtained data on Elmendorf tear strength of three different vinyl floor coverings. Table 11.5 provides the data, in grams, for two of the three vinyl floor coverings. At the 5% significance level, do the data provide sufficient evidence to conclude that the population standard deviations of tear strength differ for the two vinyl floor coverings?

Table 11.5

Brand A		Brand B	
2288	2384	2592	2384
2368	2304	2512	2432
2528	2240	2576	2112
2144	2208	2176	2288
2160	2112	2304	2752

Solution: We assume the Brand A data is in List 1 and the Brand B data is in List 2.

To begin, we construct normal probability plots (not shown) for the two samples in Table 11.5. The plots suggest that it is reasonable to presume that tear strength is normally distributed for each brand of vinyl flooring. Therefore, we can use the F-test for two population standard deviations.

Step 1: State the null and alternative hypotheses.
Let σ_1 and σ_2 denote, respectively, the population standard deviations of tear strength for Brand A and Brand B. The null and alternative hypotheses are:

H_o: $\sigma_1 = \sigma_2$ (standard deviations of tear strength are the same.)

H_a: $\sigma_1 \neq \sigma_2$ (standard deviations of tear strength are different.)

Step 2: Decide on the significance level, α.
The test is to be performed at the *5% significance level*. Thus *α = 0.05*.

Step 3: Compute the value of the test statistic.

1. Press [STAT] and arrow over to the TESTS menu.

2. Procedure D is the two sample F-test so press [ALPHA] D or arrow down to **D:2-SampFTest** and press [ENTER].

The TI-83/84 Plus has two different ways of running a F-test. The first is useful if you have the data, but not the sample standard deviations. We will use this method first.

3. Highlight Data and press [ENTER].

4. Enter L_1 for List 1 and L_2 for List 2 by pressing [2nd] L1 [ENTER] [2nd] L2.

5. Enter 1 for Freq1 and Freq2. (Recall the TI-83/84 Plus goes into Alpha mode for Frequencies so press [ALPHA] [1].)

6. Highlight the appropriate alternative hypothesis and press [ENTER]. For this example we will highlight ≠. Your screen should appear as in Figure 11.9.

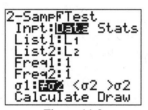
Figure 11.9

The TI-83/84 Plus has two displays for the answer. We will demonstrate both.

7. Highlight Calculate and press [ENTER]. Your screen will appear as in Figure 11.10. If you highlight Draw and press [ENTER], your screen will appear as in Figure 11.11.

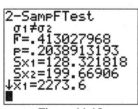
Figure 11.10

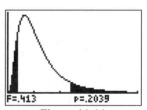
Figure 11.11

The second method of computing the test statistic and the *p*-values is useful if you do not have the data values but have the sample standard deviations.

1. Press [STAT] and arrow over to the TESTS menu.

2. Procedure D is the two sample F-test so press [ALPHA] D or arrow down to **D:2-SampFTest** and press [ENTER].

3. Highlight Stats and press [ENTER].

4. Enter the values for the standard deviations and the sample sizes. Here we will enter 128.3, 10, 199.7, and 10.

5. Highlight the appropriate alternative hypothesis and press ENTER. For this example we will highlight ≠. Your screen should appear as in Figure 11.12.

Figure 11.12

The TI-83/84 Plus has two displays for the answer. We will demonstrate both.

7. Highlight Calculate and press ENTER. Your screen will appear as in Figure 11.13. If you highlight Draw and press ENTER, your screen will appear as in Figure 11.14.

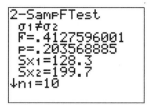
Figure 11.13

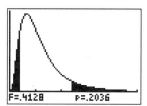
Figure 11.14

Note that both answer screens for both methods display the test statistic and the *p*-value. Here the test statistic is $F = 0.41$ when rounded to two decimal places and the *p*-value is 0.2036 when rounded to four decimal places. When using the two different methods, your answers may vary slightly and the values may differ slightly from the answers in the textbook because the TI-83/84 Plus does not round when computing the test statistic and it does not round the test statistic to compute the *p*-value. However, your final conclusion should always be the same.

Step 4: Obtain the *p*-value.
Our *p*-value is 0.2036

Step 5: If *p*-value ≤ α , reject H_0; otherwise, do not reject H_0.
Our *p*-value is 0.2036 which exceeds the specified significance level of 0.05. Therefore we do not reject the null hypothesis.

Step 6: Interpret the results of the hypothesis test.
At the 5% significance level, the data do not provide sufficient evidence to conclude that the population standard deviations of tear strength differ for the two vinyl floor coverings.

11.2 Practice Problems (continued)

Problem 11.69 One year at Arizona State University, the algebra course director decided to experiment with a new teaching method that might reduce variability in final-exam scores by eliminating lower ones. The director randomly divided the algebra students who were registered for class at 9:40 A.M. into two groups. One of the groups, called the control group, was taught the usual algebra course; the other group, called the experimental group, was taught by the new teaching method. Both classes covered the same material, took the same unit quizzes, and the same final exam at the same time. The final-exam scores 9 out of 40 possible) for the two groups are shown in Table 11.6. At the 5% significance level, do the data provide sufficient evidence to conclude that there is less variation among final-exam scores when the new teaching method is used? Assume the populations are normally distributed.

Table 11.6

Control						Experimental			
36	35	35	33	32	32	36	35	35	31
31	29	29	28	28	28	30	29	27	27
27	27	27	26	26	25	26	23	21	21
24	24	24	23	20	20	35	32	28	28
19	19	18	18	18	17	25	23	21	19
17	16	15	15	15	15				
14	11	10	9	4					

Problem 11.71 Patients who undergo chronic hemodialysis often experience severe anxiety. Videotapes of progressive relaxation exercises were shown to one group of patients and neutral videotapes to another group. Then both groups took the State-Trait Anxiety Inventory, a psychiatric questionnaire used to measure anxiety, where higher scores correspond to higher anxiety. In the paper "The Effectiveness of Progressive Relaxation in chronic Hemodialysis Patients" (*Journal of Chronic Diseases*, 35(10)), R. Alarcon et al. presented the results of the study. The data in Table 11.7 are based on those results. Do the data provide sufficient evidence to conclude that variation in anxiety-test scores differs between patients who are shown videotapes of progressive relaxation exercises and those who are shown neutral videotapes? Use a 10% significance level. Assume the populations are normally distributed.

Table 11.7

Relaxation tapes				Neutral tapes			
30	41	28	14	36	44	47	45
40	36	38	24	50	54	54	45
61	36	24	45	50	46	28	35
38	43	32	28	42	35	32	43
37	34	20	23	41	33	35	36
34	47	25	31	32	17	45	
39	14	43	40	24	46		
29	21	40					

Problem 11.81 Anthropologists are still trying to unravel the mystery of the origins of the Etruscan empire, a highly advanced Italic civilization formed around the eighth century BC in Central Italy. Were they native to the Italian peninsula or, as many aspects of their civilization suggest, did they migrate from the East by land or sea? The maximum head breadth, in millimeters, of 70 modern Italian male skulls and 84 preserved Etruscan male skulls were analyzed to help decide whether the Etruscans were native to Italy. The data can be found on WeissStats CD. At the 5% significance level, do the data provide sufficient evidence to conclude that variation in skull measurements differs between the two populations? (Data taken from N. A. Barnicot and D. R. Brothwell, "The Evaluation of Metrical Data in the Comparison of Ancient and Modern Bones." In *Medical Biology and Etruscan Origins*, G. E. W. Wolstenholme and C. M. O'Connor, eds. Little, Brown & Co.) Assume the populations are normally distributed.

The Two-Standard-Deviation F-Interval Procedure
Procedure 11.4 of Section 11.2 outlines the procedure for constructing a confidence interval for the ratio of two population standard deviations, σ_1 and σ_2. The assumptions for the **two-standard-deviation F-Interval** are the as for the two-standard-deviation F-test.

The TI-83/84 Plus does not have a built in procedure for this interval. However, the interval may be computed using the FINT program contained on the WeissStats CD.

Example 11.15 Use the sample data in example 11.14 to determine a 95% confidence interval for the ratio,

$$\frac{\sigma_1}{\sigma_2}$$ of the standard deviations of tear strength for Brand A and Brand B vinyl floor coverings.

Solution: As we discovered in example 11.14, it is reasonable to presume the assumptions are met. Similar to the built-in procedures, FINT allows you to choose between entering data as a list or just using the statistics. We will demonstrate both methods. The first is useful if you have the data values but have not computed the sample standard deviations. We will use this method first. We assume the data is in lists named BRANA and BRANB.

   1. Press PRGM and arrow down to FINT and press ENTER. prgmFINT will appear on your home screen.

   2. Press ENTER to run the program.

   3. Enter your alpha value, here 0.05, and press ENTER.

   4. Enter a 1 and press ENTER to choose DATA option.

   5. Enter the name of the first list, here BRANA, by pressing 2nd LIST, arrow to the name of the list and and press ENTER.

   6. Enter the name of the second list, here BRANB, by pressing 2nd LIST, arrow to the name of the list and and pressing ENTER.

The program will compute and display the endpoints of your confidence interval. See Figure 11.15.

Figure 11.15

Here the endpoints are 0.320 and 1.290 when rounded to three decimal places.

The second method of computing the interval is useful if you do not have the data values but have the sample standard deviations and sample sizes.

   1. After beginning the execution of the program and entering your alpha value, enter a 2 and press ENTER to choose the STATS option.

   2. Enter your degrees of freedom for the numerator, here dfn = 10-1 =9, and press ENTER.

3. Enter your degrees of freedom for the denominator, here dfd = 10-1 =9, and press ENTER.

4. Enter your sample standard deviation for sample 1, here 128.3, and press ENTER.

5. Enter your sample standard deviation for sample 2, here 199.7, and press ENTER.

The program will compute and display the endpoints of your confidence interval. See Figure 11.6.

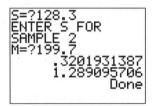

Figure 11.16

Here the endpoints are 0.320 and 1.290 when rounded to three decimal places. When using the two different methods, your answers may vary slightly and the values may vary slightly from the answers in the textbook because the TI83/84 Plus does not round anything when computing the interval.

We can be 95% confident that the ratio of the population standard deviations of Brand A and Brand B falls somewhere between 0.320 and 1.290. Notice that the confidence interval for the ratio contains 1 which means the two population standard deviations may be equal.

Problem 11.75   Use the data in problem 11.69 to obtain a 95% confidence interval for the ratio of the standard deviations of final exam scores of the two teaching methods. Interpret your result.

Problem 11.77   Use the data in problem 11.71 to obtain a 90% confidence interval for the ratio of the standard deviations of anxiety test scores of patients shown relaxation tapes and neutral tapes. Interpret your result.

Problem 11.81 (continued) Use the data in problem 11.81 to obtain a 95% confidence interval for the ratio of the standard deviations of skull measurements of modern Italian and Etruscan male skulls. Interpret your result.

CHAPTER 12
INFERENCES FOR POPULATION PROPORTIONS

LESSON 12.1   CONFIDENCE INTERVALS FOR ONE POPULATION PROPORTION

Procedure 12.1 of Section 12.1 outlines the steps for determining a **one-proportion z-interval**. The assumptions are:

1. Simple random sample

2. The number of successes, x, and the number of failures, n-x, are both 5 or greater

Example 12.3   A poll was taken of 1010 U.S. employees. The employees were asked whether they "play hooky," that is call in sick at least once a year when they simply need time to relax; 202 responded "yes." Use the TI-83/84 Plus and these data to obtain a 95% confidence interval for the proportion, p, of all U.S. employees who play hooky. Interpret your results in terms of percentages.

Solution: First we must verify that the assumptions hold. Here, $x = 202$ and $n = 1010$, so the first assumption holds. For the second, $n - x = 1010 - 202 = 808$, so it holds as well.

1. Press STAT and arrow over to the TESTS menu.

2. Press ALPHA A or arrow down to **A:1-PropZInt** and press ENTER.

3. Enter your x-value, here 202, and press ENTER.

4. Enter your n-value, here 1010, and press ENTER.

5. Enter the C-level, here 0.95, and press ENTER. Your screen should appear as in Figure 12.1.

6. Highlight Calculate and press ENTER. Your screen should appear as in Figure 12.2.

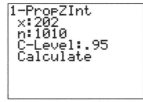   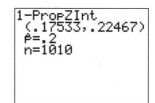

    Figure 12.1   Figure 12.2

Once again, due to rounding, the answers obtained using the TI-83/84 Plus may differ slightly from the answers in the textbook.

Here the 95% confidence interval is from 0.175 to 0.225 when rounded to three decimal places. We can be 95% confident that the percentage of all U.S. employees who play hooky is somewhere between 17.5% and 22.5%.

Notice that the one-proportion z-interval requires that you know the x and n values. Some problems will give you the value of n and the sample proportion. In those cases, multiply your value of n by the sample proportion to get your x value. If the sample proportion has been rounded then when you multiply n by the sample proportion, you will not get a whole number. Since x must be a whole number, you will need to round to the nearest whole number.

12.1 Practice Problems

Problem 12.13 A *Harris/Excite.com Internet Poll* asked Americans if states should be allowed to conduct random drug testing on elected officials? Of 21,355 respondents, 79% said "yes." Find and interpret a 95% confidence interval for the proportion of Americans who believe states should be allowed to conduct random drug testing on elected officials.

Problem 12.26 A *Reader's Digest/Gallup Survey* on the drinking habits of Americans estimated the percentage of adults across the country who drink beer, wine, or hard liquor, at least occasionally. Of the 1516 adults interviewed, 985 said they drank. Find and interpret a 95% confidence interval for the proportion of all Americans who drink beer, wine or hard liquor, at least occasionally.

Problem 12.28 From fall 1998 through mid 1999, Malaysia was the site of an encephalitis outbreak caused by the Nipah virus, a paramyxovirus that appears to spread from pigs to workers on pig farms. A reported by Goh et.al. in the *New England Journal of Medicine* (Vol. 342(17), p.1229), neurologists from the University of Malaysia found that among 94 patients infected with the Nipah virus, 30 died from encephalitis. Find and interpret a 90% confidence interval for the percentage of Malaysians infected with the Nipah virus who will die from encephalitis.

LESSON 12.2 HYPOTHESIS TESTS FOR ONE POPULATION PROPORTION

Procedure 12.2 of Section 12.2 outlines the procedure for performing the **one-proportion z-test**. The assumptions are:

1. Simple random sample
2. Both np_0 and $n(1-p_0)$ are 5 greater where p_0 is the proportion you would like to test

Example 12.6 In late January 2009, Gallup, Inc., conducted a national poll of 1053 U.S. adults that asked their views on an economic stimulus plan. The question was, "As you may know, Congress is considering a new economic stimulus package of at least 800 billion dollars. Do you favor or oppose Congress passing this legislation?" Of those sampled, 548 favored passage. At the 5% significance level, do the data provide sufficient evidence to conclude that a majority (more than 50%) of U.S. adults favored passage?

Solution: First we must check that our assumptions are met. Here our $p_0 = 0.50$ (50%) and our

n = 1053 so $np_0 = 526.5$ and $n(1-p_0) = 526.5$. Thus, the assumptions are met.

Step 1: State the null and alternative hypotheses.
Let p denote the proportion of all U.S. adults who favored passage of the economic stimulus package. Then the null and alternative hypotheses are

H_0: p = 0.50 (it is not true that a majority favor passage)
H_a: p > 0.50 (it is true that a majority favor passage)
Note that this is a right-tailed test.

Step 2: Decide on the significance level.
We are to perform the hypothesis test at the 5% significance level; so $\alpha = 0.05$.

Step 3: Compute the value of the test statistic.

1. Press [STAT] and arrow over to the TESTS menu.

2. Press [5] or arrow down to **5:1-PropZTest** and press [ENTER].

3. Enter your p_0, here, 0.5, and press [ENTER].

4. Enter your x, here 548, and press [ENTER].

5. Enter your n, here 1053, and press [ENTER].

6. Highlight your alternative hypothesis, here >p_0 and press [ENTER].   Your screen should appear as in Figure 12.3.

The TI-83/84 Plus has two displays for the answer. We will demonstrate both.

7. Highlight Calculate and press [ENTER].   Your screen should appear as in Figure 12.4. Highlight Draw and press [ENTER].   Your screen should appear as in Figure 12.5.

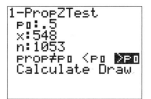

Figure 12.3

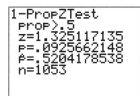

Figure 12.4

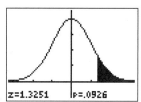

Figure 12.5

8. Note that both screens display the test statistic.   Here the test statistic is given as z= 1.33 when rounded to two decimal places. This value may differ slightly from the answers in the textbook because of differences in rounding. However, your final conclusion should be the same.

Step 4: Obtain the *p*-value.
The screen displays the *p*-value. Here the *p*-value is 0.0926 when rounded to four decimal places. This value may differ slightly from the answers in the textbook because of differences in rounding. However, your final conclusion should be the same.

Step 5: If *p*-value $\leq \alpha$, reject H$_0$; otherwise do not reject H$_0$.
Our *p*-value is 0.0926 which exceeds the significance level of 0.05. Therefore we do not reject the null hypothesis.

Step 6: Interpret the results of the hypothesis test.
At the 5% significance level, the data do not provide sufficient evidence to conclude that the majority of U.S. adults passage of the economic stimulus package.

Notice that the one-sample z-test for proportions requires that you know the x and n values. Some problems will give you the value of n and the sample proportion. In those cases, multiply your value of n by the sample proportion to get your x value. If the sample proportion has been rounded then when you multiply n by the sample proportion, you will not get a whole number. Since x must be a whole number, you will need to round to the nearest whole number.

12.2 Practice Problems

Problem 12.65 People who were born between 1978 and 1983 are sometimes classified by demographers as belonging to Generation Y. According to a recent Forrester Research survey published in *American Demographics* (Vol. 22(1), p. 12), of 850 Generation Y Web users, 459 reported using the Internet to download music. At the 5% significance level, do the data provide sufficient evidence to conclude that a majority of Generation Y Web users use the Internet to download music?

Problem 12.67 The U.S. Substance Abuse and Mental Health Services Administration conducts surveys on drug use by type of drug and age group. Results are published in *National Household Survey on Drug Abuse*. According to that publication, 13.6% of 18-25 year olds were current users of marijuana or hashish in 2000. A recent poll of 1283 randomly selected 18-25 year olds revealed that 205 currently use marijuana or hashish. At the 10% significance level, do the data provide sufficient evidence to conclude that the percentage of 18-25 year olds who currently use marijuana or hashish has changed from the 2000 percentage of 13.6%?

Problem 12.68 In 2006, 9.8% of all U.S. families had incomes below the poverty level, as reported by the Census Bureau in *Current Population Reports*. During that same year, of 400 randomly selected Wyoming families, 25 had incomes below the poverty level. At the 1% significance level, do the data provide sufficient evidence to conclude that, in 2006, the percentage of families with incomes below the poverty level was lower among those living in Wyoming thtn among all U.S. families?:

Problem 12.69 Labor Day was created by the U.S. labor movement over 100 years ago. It was subsequently adopted by most states as an official holiday. In a recent poll by the Gallup Organization, 1003 randomly selected adults were asked whether they approve of labor unions; 65% said yes. In 1936, about 72% of Americans approved of labor unions. At the 5% significance level, do the data provide sufficient evidence to conclude that the percentage of Americans who approve of labor unions now has decreased since 1936?

LESSON 12.3 INFERENCES FOR TWO POPULATION PROPORTIONS

We have studied inferences for one population proportion and now will examine inferences for comparing two population proportions. Procedure 12.3 of Section 12.3 outlines the steps for performing the **two-proportions z-test**. Procedure 12.4 of Section 12.3 outlines the steps for determining the **two-proportions z-interval**. The assumptions are:

1. Simple random samples

2. Independent samples

3. x_1, $n_1 - x_1$, x_2, and $n_2 - x_2$ are all 5 or greater. In other words, the number of successes and the number of failures are both 5 or greater in both samples.

Two-Proportions z-Test

Example 12.9 A *Zogby International* poll of 1181 adults nationwide was conducted to gauge the demand for vegetarian meals in restaurants. The study was commissioned by the Vegetarian Resource Group and was published in the *Vegetarian Journal*. In the survey, independent random samples of 747 U.S. men and 434 U.S. women were taken. Of those sampled, 276 men and 195 women said that they sometimes order a dish without meat, fish or fowl when they eat out. At the 5% significance level, do the data provide sufficient evidence to conclude that, in the United States, the percentage of men who sometimes order a dish without meat, fish, or fowl, is smaller than the percentage of women who sometimes order a dish without meat, fish or fowl?

Solution: First we will verify that the assumptions are met. $x_1 = 276$, $n_1 - x_1 = 747 - 276 = 471$, $x_2 = 195$, and $n_2 - x_2 = 434 - 195 = 239$ are all 5 or greater. In other words, the number of successes and the number of failures are both 5 or greater in both samples.

Step 1: State the null and alternative hypotheses.
Let p_1 and p_2 denote, respectively, the proportions of all U.S. men and all U.S. women who sometimes order vegetarian. Then the null and alternative hypotheses are

H_0: $p_1 = p_2$ (percentage for men is not less than that for women)

H_a: $p_1 < p_2$ (percentage for men is less than that for women)

Note that the hypothesis test is left-tailed.

Step 2: Decide on the significance level.
We are to perform the hypothesis test at the 5% significance level; so $\alpha = 0.05$.

Step 3: Compute the value of the test statistic

1. Press [STAT] and arrow over to the TESTS menu.

2. Press [6] or arrow down to **6:2-PropZTest** and press [ENTER].

3. Enter your x_1, here 276, and press [ENTER].

4. Enter your n_1, here 747, and press [ENTER].

5. Enter your x_2, here 195, and press [ENTER].

6. Enter your n_2, here 434, and press [ENTER].

7. Highlight your alternative hypothesis and press [ENTER]. We will use <p2 for our example. Your screen should appear as in Figure 12.6.

The TI-83/84 Plus has two displays for the answer. We will demonstrate both.

8. Highlight Calculate and press [ENTER]. Your screen should appear as in Figure 12.7. Highlight Draw and press [ENTER]. Your screen should appear as in Figure 12.8.

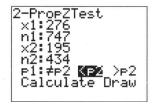

Figure 12.6

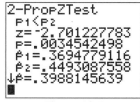

Figure 12.7

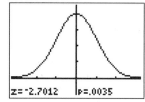

Figure 12.8

Note that both answers display the test statistic and the *p*-value. Here the test statistic is given as –2.70 when rounded to two decimal places.

Step 4: Obtain the *p*-value.
The *p*-value is 0.0035 when rounded to four decimal places. These values may differ slightly from the answers in the textbook because of differences in rounding. However, your final conclusion should be the same.

Step 5: If p-value $\leq \alpha$, reject H_0; otherwise do not reject H_0.

Our p-value is 0.0035 which is less than the specified significance level of 0.05, so we will reject the null hypothesis.

Step 6: Interpret the results of the hypothesis test.

At the 5% significance level, the data provide sufficient evidence to conclude that, in the United States, the percentage of men who sometimes order a dish without meat, fish, or fowl is smaller than the percentage of women who sometimes order a dish without meat, fish, or fowl.

Notice that the two-sample z-test for proportions requires that you know the x and n values. Some problems will give you the values of n and the sample proportions. In those cases, multiply your values of n by the appropriate sample proportions to get the respective x values.. If the sample proportions have been rounded then when you multiply n by the sample proportions, you will not get a whole number. Since x must be a whole number, you will need to round to the nearest whole number.

12.3 Practice Problems

Problem 12.89   For several years, evidence has been mounting that folic acid reduces major birth defects. An issue of *The Arizona Republic* reported on a Hungarian study that provided the strongest evidence to date. The results of the study, directed by Drs. Andrew E. Czeizel and Istvan Dudas of the National Institute of Hygiene in Budapest, were published in the paper "Prevention of the First Occurrence of Neural-Tube Defects by Periconceptional Vitamin Supplementation" (*The New England Journal of Medicine*, Vol. 327(26), p. 1832). For the study the doctors enrolled 4753 women prior to conception. The women were divided randomly into two groups. One group, consisting of 2701 women, took daily multivitamins containing 0.8 mg of folic acid; the other group, consisting of 2052 women, received only trace elements. Major birth defects occurred in 35 cases where the women took folic acid and in 47 cases where the women did not. At the 1% significance level, do the data provide sufficient evidence to conclude that women who take folic acid are at lesser risk of having children with major birth defects?

Problem 12.91   Response Insurance collects data on seat-belt usage among American drivers. Of 1000 drivers 25-34 years old, 27% said they buckle up, whereas 330 of 1100 drivers 45-64 years old said they did. At the 10% significance level, do the data suggest that there is a difference in seat-belt usage between drivers 25-34 years old and those 45-64 years old?

Problem 12.93   An adult is considered overweight if he or she has a body mass index of at least 25 kilograms/meter squared or greater. BMI is a measure that adjusts body weight for height and is based on definitions provided in the *Dietary Guideline for Americans*, published by the U.S. Department of Agriculture and the U.S. Department of Health and Human Services. Of 750 randomly selected adults whose highest degree is a Bachelors, 386 are overweight; and of 500 randomly selected adults with a graduate degree, 237 are overweight. At the 5% significance level, do the data provide sufficient evidence to conclude the percentage who are overweight is greater for those whose highest degree is a Bachelors than for those with a graduate degree

Review
Problem #14   State and local governments often poll their constituents about their views on the economy. In two polls, taken approximately 1 year apart, O'Neil Associates asked 600 Maricopa County (Arizona) residents whether they thought the state's economy would improve over the next 2 years. In the first poll, 48% said yes; in the second poll, 60% said yes. At the 1% significance level, do the data provide sufficient evidence to conclude that the percentage of Maricopa County residents who thought the state's economy would improve over the next two years was less during the time of the first poll than during the time of the second poll?

Two-Proportions z-Interval

Example 12.10   Use the vegetarian data from Example 12.9 to obtain a 90% confidence interval for the difference, $p_1 - p_2$, between the proportions of U.S. women and men who sometimes order a dish without meat, fish or fowl.

Solution: Assumptions were verified in example 12.9.

1. Press $\boxed{\text{STAT}}$ and arrow over to the TESTS menu.

2. Press $\boxed{\text{ALPHA}}$ B or arrow down to **B:2-PROPZINT** and press $\boxed{\text{ENTER}}$.

3. Enter your x_1, here 276, and press $\boxed{\text{ENTER}}$.

4. Enter your n_1, here 747, and press $\boxed{\text{ENTER}}$.

5. Enter your x_2, here 195, and press $\boxed{\text{ENTER}}$.

6. Enter your n_2, here 434, and press $\boxed{\text{ENTER}}$.

7. Enter the C-level, here 0.9, and press $\boxed{\text{ENTER}}$.   Your screen should appear as in Figure 12.9.

8. Highlight Calculate and press $\boxed{\text{ENTER}}$. Your screen should appear as in Figure 12.10.

Once again, due to rounding, the answers obtained using the TI-83/84 Plus may differ slightly from the answers in the textbook.

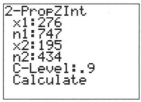

Figure 12.9

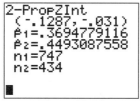

Figure 12.10

Here the 90% confidence interval is from -0.129 to -0.031 when rounded to three decimal places. We can be 90% confident that, in the United States, the percentage of men who sometimes order a dish without meat, fish or fowl is less than the percentage of women by somewhere between 3.1 and 12.9 percentage points.

Notice that the two-sample z-interval for proportions requires that you know the x and n values. Some problems will give you the values of n and the sample proportions. In those cases, multiply your values of n by the appropriate sample proportions to get the respective x values. If your answer is not a whole number, round up the nearest whole number

12.3 Practice Problems (continued)

Problem 12.95   Use the folic acid data from Problem 12.89 to obtain and interpret a 98% confidence interval for the difference, $p_1 - p_2$, between the rates of major birth defects for babies born to women who have taken folic acid and those born to women who have not.

Problem 12.97   Use the seat-belt data from Problem 12.91 to obtain and interpret a 90% confidence interval for the difference between the proportions of seat-belt users for drivers in the age groups 25-34 years and 45-64 years.

Problem 12.99   Refer to Exercise 12.93.Determine and interpret a 95% confidence interval for the difference between the percentages of adults in the two degree categories who have an above healthy weight.

Review
Problem #15   Use the poll data from Review Problem #14 to obtain and interpret a 99% confidence interval for the difference between the proportion of Maricopa County residents who thought the state's economy would improve over the next two years in the first poll and the second poll.

CHAPTER 13
CHI-SQUARE PROCEDURES

LESSON 13.1   THE CHI-SQUARE DISTRIBUTION

A variable is said to have a chi-square distribution if its distribution has a chi-square (χ^2) curve. The particular χ^2-curve is specified by stating its number of degrees of freedom, similar to the t-distribution. The TI-83/84 Plus has a built in function to find the area under a χ^2-curve with specified degrees of freedom to the left of a χ^2 value.

Example   For a χ^2-curve with df = 15, find the area under the curve to the left of 6.262.

Solution:

1. Press 2nd DISTR to access the distribution menu.

2. Arrow down to χ^2 **cdf(**   and press ENTER.

3. The format of this command is χ^2 **cdf**(lowerbound, upperbound, df). For our problem we are interested in the area from 0 to 6.262. For our problem our command is therefore χ^2 **cdf**(0, 6.262, 15). The result is shown in Figure 13.1.

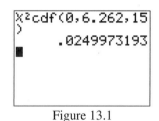

Figure 13.1

If you are looking for the area to the right of a χ^2-value, you will need to estimate ∞ as 1E99. Keep in mind that the χ^2-curve is not symmetric unlike the normal and t-curves, but the total area under the curve is equal to 1.

The TI-83/84 Plus does not have a built-in Inv χ^2 function like InvNorm. However, the program INVCHI contained in the WeissStats CD can be used to accomplish the same thing.

Example 13.1   For a χ^2-curve with df = 12, find $\chi^2_{0.025}$; that is, find the χ^2- value having area 0.025 to its right.

Solution:

1. Press PRGM and arrow down to INVCHI and press ENTER. prgmINVCHI should appear on your home screen.

2. Press ENTER to run the program.

3. Follow the prompts entering 12 for the DF, 2 for area to the right, and 0.025 for the area. The calculator will take a little while to do the computation. See Figure 13.2 for final result.

Figure 13.2

The χ^2-value having area 0.025 to its right is 23.337 when rounded to three decimal places.

13.1 Practice Problems

Problem 13.5   For a χ^2-curve with 19 degrees of freedom, determine the χ^2-value that has area
a) 0.025 to its right   b) 0.95 to its right

Problem 13.7   For a χ^2-curve with df = 10, determine
a) $\chi^2_{0.05}$ b) $\chi^2_{0.975}$

LESSON 13.2   CHI-SQUARE GOODNESS-OF-FIT TEST

This procedure can be used to perform a hypothesis test about the distribution of a qualitative (categorical) variable or a discrete quantitative variable having only a finite number of possible values. The assumptions for this test are as follows:

1. All expected frequencies are 1 or greater

2. At most 20% of the expected frequencies are less than 5

3. Simple random sample

Example 13.3   The U.S. Federal Bureau of Investigation (FBI) compiles data on crimes and crime rates and publishes the information in *Crime in the United States*. A violent crime is classified by the FBI as murder, forcible rape, robbery, or aggravated assault. Table 13.1 shows the relative frequency distribution for (reported) violent crimes for 2000. A random sample of 500 violent-crime reports from last year yielded the frequency distribution shown in Table 13.2. Do the data provide sufficient evidence to conclude that last year's distribution of violent crimes has changed from the 2000 distribution? Perform the hypothesis test at the 0.05 level of significance.

Table 13.1

Type	Rel. Freq.
Murder	0.011
Forcible Rape	0.063
Robbery	0.286
Agg. Assault	0.640
	1.000

Table 13.2

Type	Freq.
Murder	3
Forcible Rape	37
Robbery	154
Agg. Assault	306
	500

Solution: We will use the program CHIGFT contained on the WeissStats CD to compute the test statistic and the *p*-value. This program does not check the assumptions so you must do that after the test is run.

Step 1: State the null and alternative hypotheses.
H_0: Last year's violent-crime distribution is the same as the 2000 distribution.
H_a: Last year's violent-crime distribution is different from the 2000 distribution.

Step 2: Decide on the significance level.
We are to perform the hypothesis test at the 5% significance level; so $\alpha = 0.05$.

Step 3: Compute the value of the test statistic
The TI-84 Plus has a built in procedure to run a χ^2 goodness of fit test. However, the TI-83 Plus does not have a built in procedure. However, the program CHIGFT contained in the WeissStats CD will compute the test statistic and the *p*-value. We will demonstrate both solutions.

Using the TI-84 Plus

1. Enter the observed frequencies into list L1. The observed values are given in Table 13.2.

2. Enter the expected frequencies into list L2. Note: The expected frequencies are based on the total sample size. In this example, the sample size is 500. You must multiply each of the relative frequencies in Table 13.1 by 500. The expected frequencies are 5.5, 31.5, 143, and 320, respectively. Since the minimum expected frequency is 5.5 the assumptions are met.

3. Press [STAT] and arrow over to the TESTS menu. Arrow down to D: χ^2 GOF or press [ALPHA] D. Then press [ENTER].

4. Enter the observed list by pressing [2nd] L1 [ENTER].

5. Enter the observed list by pressing [2nd]

6. Enter the degrees of freedom which is defined as df = k-1 where k is the number of possible categories for the variable.   In this example, df = 4 - 1 =3. Press [ENTER]. See Figure 13.3 for the completed screen.

The TI-84 has two displays for the answer. We will demonstrate both.

7. Highlight Calculate and press [ENTER]. The screen will appear as in Figure 13.4. Highlight Draw and press [ENTER]. The screen will appear as in Figure 13.5.

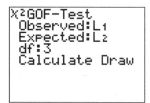

Figure 13.3

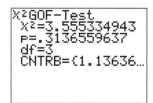

Figure 13.4

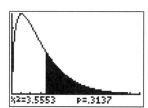

Figure 13.5

Using the TI-83 Plus

For the TI-83 Plus you must use the program CHIGFT contained in the WeissStats CD.

The CHIGFT program requires that the observed frequencies be in List 1 and the relative frequencies be in List 2. The program then computes the expected frequencies and stores them in List 3. It then computes the test statistic. Because this test is always right-tailed, the program computes the area under the curve to the right of the test statistic for the *p*-value.

To run the program:

1. Press PRGM.

2. Look for the CHIGFT name and press the number or arrow down to the CHIGFT line and press ENTER.

3. Press ENTER to run the program. The test statistic and the *p*-value will be displayed. See Figure 13.6

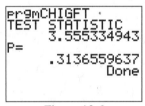

Figure 13.6

The test statistic is $\chi^2 = 3.555$.

Step 4: Obtain the *p*-value.
Here our *p*-value is 0.314 when rounded to three decimal places.

Step 5: If *p*-value $\leq \alpha$ reject H_0; otherwise do not reject H_0.
Since the *p*-value exceeds the significance level of 0.05, we do not reject H_0.

Step 6: Interpret the results of the hypothesis test.
At the 5% significance level, the data do not provide sufficient evidence to conclude that last year's violent-crime distribution differs from the 2000 distribution.

13.2 Practice Problems

Problem 13.27   The Higher Education Research Institute of the University of California, Los Angeles, publishes information on characteristics of incoming college freshmen in *The American Freshman*. In 2000, 27.7% of incoming freshmen characterized their political views as liberal, 51.9% as moderate, and 20.4% as conservative. For this year, a random sample of 500 incoming college freshmen yielded the frequency distribution shown in Table 13.3. At the 5% significance level, do the data provide sufficient evidence to conclude that this year's distribution of political views for incoming college freshmen has changed from the 2000 distribution?

Table 13.3

Political View	Freq.
Liberal	160
Moderate	246
Conservative	94

Problem 13.29   Observing that the proportion of blue M&Ms in his bowl of candy appeared to be less than that of the other colors, Ronald D. Fricker, Jr., decided to compare the color distribution in randomly chosen bags of M&Ms to the theoretical distribution reported by M&M/MARS consumer affairs. Fricker published his findings in the article "The Mysterious Case of the Blue M&Ms" (*Chance*, Vol. 9(4), pp. 19-22). The theoretical distribution is given in Table 13.4. For his study, Fricker bought three bags of M&Ms from local stores and counted the number of each color. The average number of each color in the three bags is given in Table 13.5. At the 5% significance level, do the data provide sufficient evidence to conclude that the color distribution of M&Ms differs from that reported by M&M/MARS consumer affairs?

Table 13.4

Color	Rel. Freq.
Brown	0.3
Yellow	0.2
Red	0.2
Orange	0.1
Green	0.1
Blue	0.1

Table 13.5

Color	Freq.
Brown	152
Yellow	114
Red	106
Orange	51
Green	43
Blue	43

Problem 13.31   A gambler thought a die was loaded (i.e. the six numbers were not equally likely.) To test his suspicion, he rolled the die 150 times and obtained the data shown in Table 13.6. Do the data provide sufficient evidence to conclude that the die was loaded? Perform the hypothesis test at the 0.05 level of significance.

Table 13.6

Number	1	2	3	4	5	6
Frequency	23	26	23	21	31	26

LESSON 13.3 CONTINGENCY TABLES; ASSOCIATION

A **contingency table** or **two-way table** shows the frequency distribution of sample data that is simultaneously grouped into cells by two variables of a population. A contingency table is also known as a **cross-tabulation table** or **cross-table.**

Next, we need to discuss the concept of **association** between two variables. We do so for variables that are either categorical or quantitative with only finitely many possible values. We say that two variables are associated (or that an **association** exists between the two variables) if the conditional distributions of one variable given the other variable are not identical. Roughly speaking, we say that two variables are associated if knowing the value of one of the variables gives information about the other variable.

LESSON 13.4 CHI-SQUARE INDEPENDENCE TEST

The chi-square independence test allows us to decide whether or not an association exists between two variables.

The assumptions for this test are:

1. All expected frequencies are at least 1

2. At most 20% of the expected frequencies are less than 5

3. Simple random sample

The TI-83/84 Plus will compute the expected frequencies, the test statistic and the *p*-value for this test. However, after the test is run, the user must check the assumptions.

Example 13.10   A national survey was conducted to obtain information on the alcohol consumption patterns of U.S. adults by marital status. A random sample of 1772 residents, 18 years old and over, yielded the data displayed in Table 13.7. At the 5% significance level, do the data provide sufficient evidence to conclude that an association exists between marital status and alcohol consumption?

Table 13.7

	Abstain	1-60	Over 60	Total
Single	67	213	74	354
Married	411	633	129	1173
Widowed	85	51	7	143
Divorced	27	60	15	102
Total	590	957	225	1772

Solution:

Step 1: State the null and alternative hypotheses.
H_0: Marital status and alcohol consumption are not associated.
H_a: Marital status and alcohol consumption are associated.

Step 2: Decide on the significance level.
The test is to be performed at the 5% significance level. Thus $\alpha = 0.05$.

Step 3: Compute the value of the test statistic
In order to run the chi-square test, the table of observed frequencies must be placed in a matrix in the calculator. Only the observed frequencies are entered, not the totals.

1. Press 2nd MATRX, arrow over to EDIT, and press ENTER.

2. Enter the number of rows, here 4, and press ENTER.

3. Enter the number of columns, here 3, and press ENTER.

4. Begin entering your data values going across the rows and pressing ENTER after <u>each</u> entry. Be sure you do <u>not</u> enter the totals, and that you press ENTER after the last entry. See Figure 13.7 for final screen.

5. Exit the matrix edit mode by pressing 2nd QUIT.

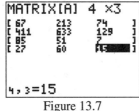
Figure 13.7

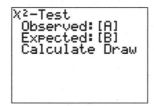
Figure 13.8

Now we are ready to run the test. The TI-83/84 Plus has several matrices and we have used matrix A for this example.

6. Press STAT and arrow over to TESTS.

7. Press ALPHA C or arrow down to **C:** χ^2-Test and press ENTER.

8. For Observed enter the matrix containing your observed frequencies. Here that is matrix A so press [2nd] MATRX [1] or press [2nd] MATRX [ENTER].

9. Your calculator will return to the chi-square test menu. Press [ENTER].

10. Enter the matrix where you would like your expected frequencies stored. Here we will use matrix B. Press [2nd] MATRX [2].

11. Your calculator will return to the chi-square test menu. Press [ENTER]. See Figure 13.8.

The TI-83/84 Plus has two displays for the answer. We will demonstrate both.

12. Highlight Calculate and press [ENTER]. Your screen will appear as in Figure 13.9. If you highlight Draw and press [ENTER], your screen will appear as in Figure 13.10.

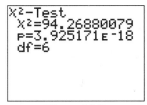

Figure 13.9

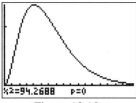

Figure 13.10

Our test statistic is 94.269 when rounded to three decimal places and the *p*-value is approx 0. To check the assumptions, we must view the matrix that holds our expected frequencies.

13. From the home screen, press [2nd] MATRX and then the number of the matrix where your expected frequencies are stored. We stored the values to matrix B so press [2].

14. Press [ENTER].

15. The matrix will appear on your screen. However, due to decimal places, not all of the matrix may appear at one time. Use your left and right arrow keys to scroll back and forth so you can see all the entries. See Figures 13.11, 13.12, and 13.13 for examples of the screen.

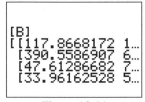

Figure 13.11

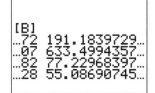

Figure 13.12

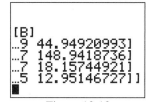

Figure 13.13

If we round these values to one decimal place, we would get Table 13.8 for the expected frequencies.

Table 13.8

	Abstain	1-60	Over 60	Total
Single	117.9	191.2	44.9	354
Married	390.6	633.5	148.9	1173
Widowed	47.6	77.2	18.2	143
Divorced	34.0	55.1	13.0	102
Total	590	957	225	1772

All of these values are greater than 5 so both assumptions are met. Therefore our test is valid.

Step 4: Obtain the *p*-value.
Here our *p*-value is 3.93E-18.

Step 5: If *p*-value $\leq \alpha$, reject H$_0$; otherwise do not reject H$_0$.
Given that our *p*-value from above was approximately 0, we reject H$_o$.

Step 6: Interpret the results of the hypothesis test.
At the 5% significance level, the data provide sufficient evidence to conclude that there is an association between marital status and alcohol consumption.

13.4 Practice Problems

Problem 13.71   In Gene Siskel and Roger Ebert's show *Sneak Preview*, the two Chicago movie critics reviewed the week's new movie releases and then rated them thumbs up (positive), mixed, or thumbs down (negative). These two critics often saw the merits of a movie differently. But, in general, were the ratings given by Siskel and Ebert associated? The answer to the question was the focus of the paper "Evaluating Agreement and Disagreement Among Movie Reviewers" by Alan Agresti and Larry Winner that appeared in *Chance* (Vol. 10(2), pp. 10-14). Table 13.9 summarizes the ratings by Siskel and Ebert for 160 movies. At the 1% significance level, do the data provide sufficient evidence to conclude that an association exists between the ratings of Siskel and Ebert?

Table 13.9

		Ebert			
		Thumbs down	Mixed	Thumbs up	Total
Siskel	Thumbs down	24	8	13	45
	Mixed	8	13	11	32
	Thumbs up	10	9	64	83
	Total	42	30	88	160

Problem 13.73   M. Stuart et al. studied various aspects of grade-school children and their mothers and reported their findings in the article "Learning to Read at Home and at School" (*British Journal of Educational Psychology*, 68(1), pp. 3-14),. The researchers gave a questionnaire to parents of 66 children in kindergarten through second grade. Two social-class groups, middle and working, were identified based on the mother's occupation. One of the questions dealt with whether the parents played "I Spy" games with their children. The data in Table 13.10 were obtained. At the 5% significance level, do the data provide sufficient evidence to conclude that social-class and the frequency of playing "I Spy" games are associated?

Table 13.10

		Frequency of Games		
		Never	Sometimes	Often
Social	Middle	2	8	22
Class	Working	11	10	13

Problem 13.75 The American Bar Foundation publishes information on the characteristics of lawyers in *The Lawyer Statistical Report*. Table 13.11 cross classifies 307 randomly selected U.S. lawyers by status in practice and the size of the city in which they practice. At the 5% significance level, do the data provide sufficient evidence to conclude that size of city and status in practice are statistically dependent for U.S. lawyers?

Table 13.11

			Size		
		Less than 250,000	250,000-499,999	500,000 or more	Total
Status	Govern-ment	12	4	14	30
	Judicial	8	1	2	11
	Private Practice	122	31	69	222
	Salaried	19	7	18	44
	Total	161	43	103	307

LESSON 13.5   CHI-SQUARE HOMOGENEITY TEST

The purpose of a **chi-square homogeneity test** is to compare the distributions of a variable of two or more populations. As a special case, it can be used to decide whether a difference exists among two or more population proportions. For a chi-square homogeneity test, the null hypothesis is that the distributions of the variable are the same for all the populations, and the alternative hypothesis is that the distributions of the variable are not all the same (i.e., the distributions differ for at least two of the populations). The assumptions for this test are:

1. All expected frequencies are at least 1

2. At most 20% of the expected frequencies are less than 5

3. Simple random samples

4. Independent samples

Although the context of and assumptions for the chi-square homogeneity test differ from those of the chi-square independence test, the steps for carrying out the two tests are the same. In particular, the test statistics for the two tests are identical. As with a chi-square independence test, the observed frequencies for a chi-square homogeneity test are arranged in a contingency table. Moreover, the expected frequencies are computed in the same way.

Example 13.13 The Organization for Economic Cooperation and Development compiles information on unemployment rates of selected countries and publishes its findings in *Main Economic Indicators*. Independent simple random samples from the civilian labor forces of the five Scandinavian countries—Denmark, Norway, Sweden, Finland, and Iceland—yielded the data in Table 13.12 on employment status. At the 5% significance level, do the data provide sufficient evidence to conclude that a difference exists in the unemployment rates of the five Scandinavian countries?

Table 13.12

	Status		
	Unemployed	Employed	Total
Denmark	12	309	321
Norway	7	265	272
Sweden	32	498	530
Finland	21	286	307
Iceland	1	69	70
Total	73	1427	1500

Solution: We will use the chi-square homogeneity test in order to determine if there is sufficient evidence to conclude a difference exists in the unemployment rates of the five Scandinavian countries.

Step 1: State the null and alternative hypotheses.
Let p_1, p_2, p_3, p_4, and p_5 denote the population proportions of the unemployed people in the civilian labor forces of Denmark, Norway, Sweden, Finland, and Iceland, respectively.

H_0: $p_1 = p_2 = p_3 = p_4 = p_5$ (unemployment rates are equal).
H_a: Not all the unemployment rates are equal

Step 2: Decide on the significance level, α.
The test is to be performed at the *5% significance level. Thus α = 0.05.*

Step 3: Compute the value of the test statistic
In order to run the chi-square test, the table of observed frequencies must be placed in a matrix in the calculator. Only the observed frequencies are entered, not the totals.

1. Press 2nd MATRX, arrow over to EDIT, and press ENTER.

2. Enter the number of rows, here 4, and press ENTER.

3. Enter the number of columns, here 3, and press ENTER.

4. Begin entering your data values going across the rows and pressing ENTER after each entry. Be sure you do **not** enter the totals, and that you press ENTER after the last entry. See Figure 13.4 for final screen.

5. Exit the matrix edit mode by pressing 2nd QUIT.

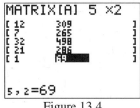

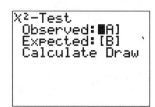

Figure 13.4   Figure 13.5

Now we are ready to run the test. The TI-83/84 Plus has several matrices and we have used matrix A for this example.

6. Press STAT and arrow over to TESTS.

7. Press ALPHA C or arrow down to **C:** χ^2-Test and press ENTER.

8. For Observed enter the matrix containing your observed frequencies. Here that is matrix A so press 2nd MATRX 1 or press 2nd MATRX ENTER.

9. Your calculator will return to the chi-square test menu. Press ENTER.

10. Enter the matrix where you would like your expected frequencies stored. Here we will use matrix B. Press 2nd MATRX 2.

11. Your calculator will return to the chi-square test menu. Press ENTER. See Figure 13.5. The TI-83/84 Plus has two displays for the answer. We will demonstrate both.

12. Highlight Calculate and press ENTER. Your screen will appear as in Figure 13.6. If you highlight Draw and press ENTER, your screen will appear as in Figure 13.7.

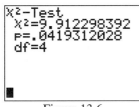

Figure 13.6

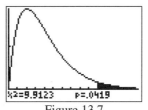

Figure 13.7

The χ^2- value of 9.912 is the test statistic. It has a χ^2-distribution with 4 degrees of freedom.

Step 4: Obtain the *p*-value.
  The *p*-value is 0.042 rounded to 3 decimal places.

Step 5: If *p*-value $\leq \alpha$, reject H_0; otherwise, do not reject H_0.
  Because the *p*-value is less than the specified significance level of 0.05, we reject H_0.

Step 6: Interpret the results of the hypothesis test.
  At the 5% significance level, the data provide sufficient evidence to conclude that a difference exists in the unemployment rates of the five Scandinavian countries.

13.5 Practice Problems

Problem 13.92   Prior to the 2008 election, the Quinnipiac University Poll asked a sample of U.S. residents, "If Barack Obama is elected President, do you think the economy will get better, get worse or stay about the same?" This problem is based on the results of that poll. Independent simple random samples of 500 residents each in red (predominantly Republican), blue (predominantly Democratic), and purple (mixed) states responded to the aforementioned question as follows in Table 13.13   At the 5% significance level, do the data provide sufficient evidence to conclude that the residents of the red, blue, and purple states are nonhomogeneous with respect to their view?

Table 13.13

		State Classification			
		Red	Blue	Purple	Total
	Get Better	169	191	196	556
View	Get Worse	129	89	106	324
	Stay Same	149	178	161	488
	Don't Know	53	42	37	132
	Total	500	500	500	1500

Problem 13.96   From two *USA Today/Gallup* polls, we found information about Americans' approval of government bailouts to two of the Big Three U.S. automakers. The question asked was, "Do you approve or disapprove of the federal loans given to General Motors and Chrysler last year to help them avoid bankruptcy?" In a February 2009 poll of 1007 national adults, 41% said they approved, and in a March 2009 poll of 1007 national adults, 39% said they approved. At the 5% significance level, do the data provide sufficient evidence to conclude that a difference exists in the approval percentages of all U.S. adults between the two months? Use the chi-square homogeneity test.

CHAPTER 14
DESCRIPTIVE METHODS IN REGRESSION AND CORRELATION

LESSON 14.1   LINEAR EQUATIONS WITH ONE INDEPENDENT VARIABLE

The general form of a linear equation with one independent variable can be written as $y = b_0 + b_1 x$ where b_0 and b_1 are constants (fixed numbers), x is the independent variable, and y is the dependent variable.

The TI-83/84 Plus is capable of graphing a linear equation.

Example 14.1 CJ^2 Business Services does word processing as one of its basic functions. The total cost, y, of a job that takes x hours is y = 25 +20x.   Graph the cost equation using the TI-83/84 Plus.

Solution: We will use the Y= screen to plot linear equations.

1. Press Y= to enter the Y= screen. Your screen should appear as in Figure 14.1. Note that your stat plots should be turned off, meaning the plot icons are not highlighted. If any of your plots are highlighted, arrow up to the plot so it is highlighted by your cursor and press enter. Arrow off the plot and the plot icon will no longer be highlighted.

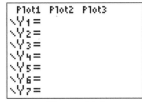
Figure 14.1

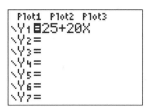
Figure 14.2

2. Position your cursor at Y_1 = as in Figure 14.1 and enter your equation by pressing 25 + 20 X,T,Θ,n ENTER. See Figure 14.2.

Now we must set the appropriate window for our plot. We must examine what sort of y-values we get based on the x-values we expect to use. Because we are using x to represent the number of hours required to complete the job, it is reasonable to expect values above zero. Because y is the cost based on hours, we can expect some fairly large numbers. If the time is 5 hours, the cost is $125.

3. Press WINDOW to access the window setup screen. Enter Xmin = 0, Xmax = 25, Xscl = 5, Ymin = 0, Ymax = 500 and Yscl = 100. We will keep the Xres = 1. Your screen should appear as in Figure 14.3.

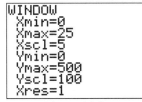
Figure 14.3

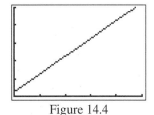
Figure 14.4

4. Press GRAPH. Your screen should appear as in Figure 14.4. The graph is the graph of your equation and you may use the TRACE key to trace along it.

The equation we graphed slopes upward which we would expect given a positive slope.

LESSON 14.2   THE REGRESSION EQUATION

Real life applications are sometimes not so simple that one variable can be predicted exactly in terms of another variable. Often we must be content with rough predictions. One method is to compute a linear regression equation. In order for this to be a reasonable estimate, we must first verify that the data is scattered about a straight line. This is usually done by creating a scatterplot.

Example 14.4   Table 14.1 displays data on age and price for a sample of cars of a particular make and model. We will refer to the car as the Orion. Ages are in years; prices are in hundreds of dollars, rounded to the nearest hundred dollars. Create a scatterplot to see if it is appropriate to use linear regression on this problem.

Table 14.1

Car	Age (yrs) x	Price ($100s) y
1	5	85
2	4	103
3	6	70
4	5	82
5	5	89
6	5	98
7	6	66
8	6	95
9	2	169
10	7	70
11	7	48

Solution: We will assume the x values are in List 1 and the y values are in List 2.

Before attempting to graph any statistical plot, be sure that your function plot screen is clear, or the plots are turned off.

1. Enter the stat plot screen by pressing [2nd] STATPLOT. Your screen will appear as in Figure 14.5.

2. Enter the set-up screen for plot 1 by pressing [ENTER]. Your screen should appear similar to Figure 14.6.

Figure 14.5

Figure 14.6

3. Turn Plot 1 on by highlighting On and pressing [ENTER].

4. Select the scatterplot by highlighting the scatterplot icon ⋅⋅̈ (the first one in the Type list) and pressing [ENTER].

5. Arrow down to Xlist and press [2nd] L1 for List 1. Arrow down to Ylist and press [2nd] L2 for List 2.

6. Select the mark you would like to represent the points on the graph. For this example we will choose the box. Highlight the box and press [ENTER]. See Figure 14.7 for your final screen.

7. Press ZOOM 9 for ZoomStat and the graph will appear. See Figure 14.8. You may choose to adjust your window for a different view. Press WINDOW and change your values. Then press GRAPH.

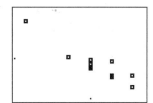

| Figure 14.7 | Figure 14.8 |

14.2 Practice Problems

Problem 14.51 The *Kelley Blue Book* provides information on wholesale and retail prices of cars. Table 14.2 provides age and price data for 10 randomly selected Corvettes between 1 and 6 years old. Ages are given in years and prices are given in hundreds of dollars. Create a scatterplot to see if it is appropriate to use linear regression on this data.

Table 14.2	Age	6	6	6	2	2	5	4	5	1	4
	Price	270	260	275	405	364	293	335	308	405	305

Problem 14.53 Plants emit gases that trigger the ripening of fruit, attract pollinators, and cue other physiological responses. N. G. Agelopolous, K. Chamberlein, and J.A. Pickett examined factors that affect the emission of volatile compounds by the potato plant *Solanum tubersom* and published their findings in the *Journal of Chemical Ecology* (Vol. 26(2), pp. 497-511). The volatile compounds analyzed were hydrocarbons that are used by other plants and animals. Table 14.3 contains data on plant weight (x) in grams, and quantity of volatile compounds emitted (y), in hundreds of nanograms, for 11 potato plants. Create a scatterplot to see if it is appropriate to use linear regression on this data.

Table 14.3	X	57	85	57	65	52	67	62	80	77	53	68
	Y	8.0	22.0	10.5	22.5	12.0	11.5	7.5	13.0	16.5	21.0	12.0

Problem 14.63 In the paper "Mating System and Sex Allocation in the Gregarious Parasitoid *Cotesia glomerata*" (*Animal Behaviour*, Vol. 66, pp. 259-264), H. Gu and S. Dorn reported on various aspects of the mating system and sex allocation strategy of the wasp *C. glomerata*. One part of the study involved the investigation of the percentage of male wasps dispersing before mating in relation to the brood sex ratio (proportion of males). The data in Table 14.4 were obtained by the researchers. Create a scatterplot to see if it is appropriate to use linear regression on this data.

Table 14.4	Sex ratio	% dispersing	Sex ratio	% dispersing
	0.143	24.9	0.657	27.0
	0.106	29.9	0.667	32.9
	0.157	27.9	0.623	38.0
	0.294	12.8	0.680	36.9
	0.387	20.9	0.722	37.0
	0.446	18.9	0.737	33.0
	0.550	20.9	0.758	34.0
	0.455	21.8	0.787	36.0
	0.447	23.9	0.810	42.0
	0.502	27.9	0.836	42.0
	0.554	29.0	0.900	53.3

Problem 14.71 Does a higher state per capita income equate to a higher per capita beer consumption? Data downloaded from *The Beer Institute Online* on per capita income, in dollars, and per capita beer consumption, in gallons, for the 50 states and Washington, D.C., are provided on the WeissStats CD. Create a scatterplot to see if it is appropriate to use linear regression on this data.

Problem 14.105 J. Greene and J. Touchstone conducted a study on the relationship between the estriol levels of pregnant women and the birth weights of their children. Their findings, "Urinary Tract Estriol: An Index of Placental Function," were published in the *American Journal of Obstetrics and Gynecology* (Vol. 85(1), pp. 1-9). The data from the study are provided on the WeissStats CD, where estriol levels are in mg/24 hr and birth weights are in hectograms. Create a scatterplot to see if it is appropriate to use linear regression on this data.

Review
Problem #14 Graduation rate-the percentage of entering freshmen, attending full time, that graduate within 5 years-and what influences it have become a concern in U.S. colleges and universities. *U.S. News and World Report's* "College Guide" provides data on graduation rates for colleges and universities as a function of the percentage of freshmen in the top 10% of their high-school class, total spending per student, and student-to-faculty ratio. A random sample of 10 universities gave the data in Table 14.5 on student-to-faculty ratio (S/F ratio) and graduation rate (grad rate). Create a scatterplot to see if it is appropriate to use linear regression on this data.

Table 14.5

S/F ratio X	Grad rate y
16	45
20	55
17	70
19	50
22	47
17	46
17	50
17	66
10	26
18	60

Once it has been determined that the data are roughly scattered about a straight line, the problem now becomes one of picking the line that best fits the data. The method employed is called the **least-squares criterion**. The line of best fit is called the **regression line** and the equation is called the **regression equation**.

Definition 14.3: Notation Used in Regression and Correlation

We define S_{xx}, S_{xy}, and S_{yy} by

$$S_{xx} = \sum (x - \bar{x})^2 \qquad S_{xy} = \sum (x - \bar{x})(y - \bar{y}) \qquad S_{yy} = \sum (y - \bar{y})^2$$

For hand computations, these quantities are most easily obtained by using the following computing formulas.

$$S_{xx} = \sum x^2 - \left(\sum x \right)^2 / n$$

$$S_{xy} = \sum xy - \left(\sum x \right)\left(\sum y \right) / n$$

$$S_{yy} = \sum y^2 - \left(\sum y \right)^2 / n$$

Example 14.8   Table 14.1 displays data on age and price for a sample of 11 Orions. Use the TI-83/84 Plus to compute S_{xx}, S_{xy}, and S_{yy}.

Solution: We will assume that the ages are in List 1 and the prices are in List 2. The TI-83/84 Plus will compute the sums of the lists for us and then we can enter them in the formulas.

1. Press STAT and arrow over to the CALC menu.

2. Press 2 or arrow to 2 and press ENTER. **2-Var Stats** will be displayed on your home screen. Enter the names of the lists you want to sum, here List 1 and List 2, by pressing 2nd L1 , 2nd L2 ENTER. Your screen will appear as in Figure 14.9.

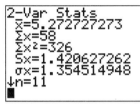

Figure 14.9

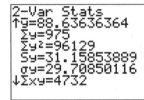

Figure 14.10

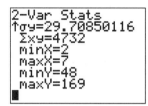

Figure 14.11

Notice the small arrow in the left bottom corner. This indicates that by pressing your down arrow key, you can obtain more information. That information is shown in Figures 14.10 and 14.11.

3. Write down the sums you need for the formulas.

4. To compute S_{xx} type $326 - (58)^2 / 11$ and press ENTER.

5. To compute S_{xy} type $4732 - (58)(975)/11$ and press ENTER.

6. To compute S_{yy} type $96129 - (975)^2 / 11$ and press ENTER. Figure 14.12 displays the results.

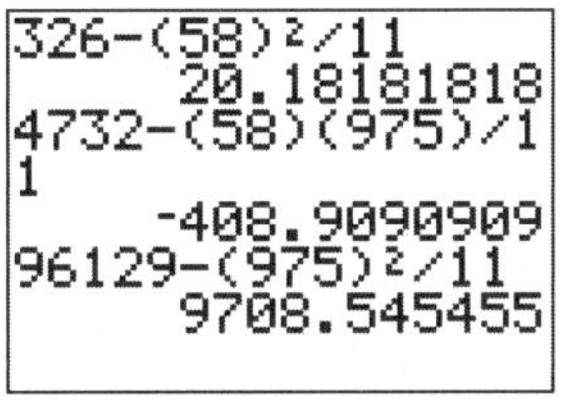

```
326-(58)²/11
            20.18181818
4732-(58)(975)/1
1
           -408.9090909
96129-(975)²/11
            9708.545455
```

Figure 14.12

Formula 14.1: The regression equation for a set of n data points is $\hat{y} = b_0 + b_1 x$, where

$$b_1 = \frac{S_{xy}}{S_{xx}} \quad \text{and} \quad b_0 = \frac{1}{n}\left(\sum y - b_1 \sum x\right) = \overline{y} - b_1 \overline{x}$$

The TI-83/84 Plus is capable of computing the regression equation directly and using that equation, can predict a y-value for any given x-value.

Example 14.4 (continued)   Table 14.1 displays data on age and price for a sample of 11 Orions.

a. Use the TI-83/84 Plus to compute the regression equation for the data.

b. Graph the regression equation and the data points.

c. Use the regression equation and the TI-83/84 Plus to predict the price of a 3-year-old   Orion and a 4-year-old Orion.

Solution: We will assume the ages are in List 1 and the prices are in List 2.

1. Press STAT and arrow over to the CALC menu.

Notice that the TI-83/84 Plus has two different linear regressions **4:LinReg(ax + b)** and **8:LinReg(a + bx)**. Number 8 most closely matches the statistics form so we will use it in this example. However, both are acceptable.

2. Press 8 or arrow down to **8:LinReg(a + bx)** and press ENTER. **LinReg (a+bx)** will appear at the top of your home screen.

3. The format for this command is LinReg(a+bx) xlist, ylist, Y-vars where the regression equation should be stored. Here we are using Lists 1 and 2 so press 2nd L1 , 2nd L2 for the lists. To get the Y where the equation will be stored, press , VARS arrow to Y-vars, press ENTER and then ENTER again to choose Y_1. Press ENTER. Your screen will appear as in Figure 14.13 or Figure 14.14.

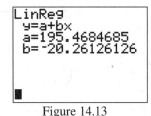
Figure 14.13

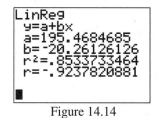
Figure 14.14

Your value for the slope is -20.26 and for the y-intercept is 195.47 when rounded to two decimal places.

b. Graph the regression equation and the data points.

1. Set up the scatterplot as we did in Example 14.2.

2. Press Y= to view the function plot screen. It should appear as in Figure 14.15. The regression equation is in Y_1 and Plot 1 is highlighted meaning it is turned on.

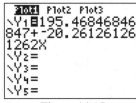
Figure 14.15

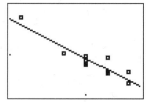
Figure 14.16

3. Press ZOOM 9 and the graph will appear as in Figure 14.16.

c. Use the regression equation and the TI-83/84 Plus to predict the price of a 3-year-old Orion and a 4-year-old Orion.

    1. From the home screen, with the regression equation still stored in Y_1, press [VARS] arrow over to Y-Vars, and press [1] or [ENTER] followed by [1] or [ENTER] to get Y_1.

    2. The format we want is function notation, so press [(] [3] [)] [ENTER] to find the price of a 3-year-old Orion.

    3. Repeat steps 1 and 2 substituting a 4 for the 3 to get the price of a 4-year-old Orion. The answers are displayed below in Figure 14.17.

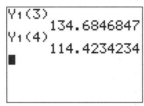

Figure 14.17

14.2 Practice Problems (continued)

Problem 14.51(continued) Table 14.2 displays data on the age and price for a sample of Corvettes.

    a. Use the TI-83/84 Plus to compute the regression equation for the data.

    b. Graph the regression equation and the data points.

    c. Use the regression equation and the TI-83/84 Plus to predict the price of a 3 year old Corvette.

Problem 14.53 (continued) Table 14.3 displays data on plant weight and quantity of volatile compounds emitted for potato plants.

    a. Use the TI-83/84 Plus to compute the regression equation for the data.

    b. Graph the regression equation and the data points.

    c. Use the regression equation and the TI-83/84 Plus to predict the quantity of volatile compounds emitted if the plant weighs 60 grams.

Problem 14.63 (continued) Table 14.4 displays data on the sex ratio and percentage of wasps dispersing.

    a. Use the TI-83/84 Plus to compute the regression equation for the data.

    b. Graph the regression equation and the data points.

    c. Use the regression equation and the TI-83/84 Plus to predict the percentage of wasps dispersing if the proportion of males is 50%.

Problem 14.71 (continued)   Data downloaded from *The Beer Institute Online* on per capita income and per capita beer consumption are provided on the WeissStats CD.

a. Use the TI-83/84 Plus to compute the regression equation for the data.

b. Graph the regression equation and the data points.

c. Use the regression equation and the TI-83/84 Plus to predict the per capita beer consumption if the per capita income is $21,000.

Problem 14.105 (continued) Data on estriol levels of pregnant women and the birth weight of their children are provided on the WeissStats CD.

a. Use the TI-83/84 Plus to compute the regression equation for the data.

b. Graph the regression equation and the data points.

c. Use the regression equation and the TI-83/84 Plus to predict the birth weight of a child whose mother had an estriol level of 13.

Review Problem #14   Table 14.5 displays data on student-to-faculty ratio and graduation rate.

a. Use the TI-83/84 Plus to compute the regression equation for the data.

b. Graph the regression equation and the data points.

c. Use the regression equation and the TI-83/84 Plus to predict the graduation rate of a university with a student-to-faculty ratio of 15.

LESSON 14.3 THE COEFFICIENT OF DETERMINATION

One way to evaluate the utility of a regression equation for making predictions is to determine the percentage of variation in the observed values of the response variables that is explained by the regression. This is called the **coefficient of determination**. To define it, we must first look at several other definitions.

Definition 14.5: Sums of Squares in Regression

Total sum of squares, SST: The variation in the observed values of the response variable.

$$SST = \sum (y - \overline{y})^2$$

Regression sum of squares, SSR: The variation in the observed values of the response variable that is explained by the regression.

$$SSR = \sum (\hat{y} - \overline{y})^2$$

Error sum of squares, SSE: the variation in the observed values of the response variable that is not explained by the regression.

$$SSE = \sum (y - \hat{y})^2$$

For hand computations, these formulas are usually computed using the following formulas:

$$SST = S_{yy} \qquad SSR = S_{xy}^2 / S_{xx} \qquad SSE = S_{yy} - S_{xy}^2 / S_{xx}$$

If the values of S_{xx}, S_{xy} and S_{yy} are known, these computations are easy. If not, the TI-83/84 Plus may be used with the defining formulas.

Example 14.8   Refer back to the data on the age and price of Orions contained in Table 14.1 and compute SST, SSR and SSE using the defining formulas.

Solution: We will assume the ages are in List 1, the prices are in List 2 and the regression equation is in Y_1.

We will compute SST first. Therefore we will use the formula $SST = \sum (y - \bar{y})^2$.

To use this formula we must first compute the mean of the y-values.

1. Press STAT arrow over to CALC and press ENTER. **1-Var Stats** will appear on your home screen.

2. Press 2nd L2 to enter List 2 and press ENTER. The mean shown will be the mean of the y-values. Here it is 88.64. See Figure 14.18.

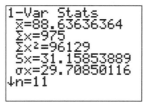

Figure 14.18

Now we must compute the values of the y's minus the mean of the y's and square them. This can be done on the home screen by manipulating lists and storing them.

3. Press (2nd L2 – 88.64) x^2 STO▶ 2nd L3 ENTER. This will take List 2 (the y's) subtract the mean, square the result and store it in List 3. Your screen will appear as in Figure 14.19.

Now we must sum this result.

4. Press 2nd LIST arrow over to MATH and press 5 for **5:sum(**. The command will appear on your home screen.

5. Enter the name of the list you wish to sum. Here that is List 3 so press 2nd L3) ENTER. Your screen will display the sum as in Figure 14.20.

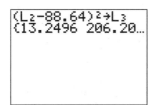

Figure 14.19

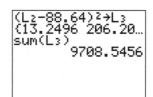

Figure 14.20

The value of SST is 9708.5 when rounded to one decimal place.

To compute SSR we use the formula $SSR = \sum (\hat{y} - \bar{y})^2$ and so we need to find the values of the predicted y's.

1. From the home screen, with the regression equation still stored in Y_1, press VARS arrow over to Y-Vars, and press 1 or ENTER followed by 1 or ENTER to get Y_1.

2. The format we want is function notation, so press (2nd L1) STO▸ 2nd L4 to find the predicted prices of Orions and store these values in List 4. See Figure 14.21.

Figure 14.21

3. Press (2nd L4 − π 88.64) x^2 STO▸ 2nd L5 ENTER. This will take List 4 (the predicted y's) subtract the mean, square the result and store it in List 4. Your screen will appear as in Figure 14.22.

Now we must sum this result.

4. Press 2nd LIST arrow over to MATH and press 5 for **5:sum(**. The command will appear on your home screen.

5. Enter the name of the list you wish to sum. Here that is List 5 so press 2nd L5) ENTER. Your screen will display the sum as in Figure 14.23.

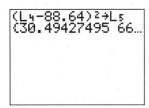

Figure 14.22

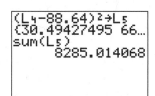

Figure 14.23

The value of SSR is 8285.0 when rounded to one decimal place.

Using the regression identity SST= SSR + SSE and solving for SSE we find it to be 1423.5

14.3 Practice Problems

Problem 14.89   Refer back to the age and price data of Corvettes in Table 14.2 and compute SST, SSR, and SSE using the defining formulas.

Problem 14.91   Refer back to the plant weight and quantity of volatile compounds emitted for potato plants in Table 14.3 and compute SST, SSR, and SSE using the defining formulas.

Problem 14.101   Refer back to the per capita income and per capita beer consumption data provided on the WeissStats CD and compute SST, SSR, and SSE using the defining formulas.

Problem 14.105   Refer back to the estriol levels of pregnant women and birth weights data provided on the

WeissStats CD and compute SST, SSR, and SSE using the defining formulas.

Review
Problem #15

Refer back to the student-to-faculty ratio and graduation rate data in Table 14.5 and compute SST, SSR, and SSE using the defining formulas.

Definition 14.6: The **coefficient of determination**, r^2, is the proportion of variation in the observed values of the response variable that is explained by the regression. We have

$$r^2 = \frac{SSR}{SST}.$$

The coefficient of determination always lies between 0 and 1 and is a descriptive measure of the utility of the regression equation for making predictions. Values of r^2 near 0 indicate that the regression equation is not very useful for making predictions, whereas values of r^2 near 1 indicate that the regression equation is extremely useful for making predictions.

The TI-83/84 Plus can be used to compute the coefficient of determination and the linear correlation coefficient without computing SSR, SST, S_{xx}, S_{xy}, or S_{yy}.

Example 14.6   Compute the coefficient of determination, r^2, for the age and price of Orions data given in Table 14.1.

Solution: The default calculator setting is to not show r^2 when linear regression is done. Therefore we must first set our calculator to show it.

1. Press [2nd] followed by [0] to access the **CATALOG**. See Figure 14.24.

   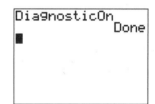

Figure 14.24   Figure 14.25

2. Press [x⁻¹] (for the d-listings), arrow down to **DiagnosticOn** and press [ENTER]. Your calculator will return to the home screen with DiagnosticOn displayed.

3. Press [ENTER] and the calculator will display the word Done. See Figure 14.25.

We will assume the ages are in List 1 and the prices are in List 2.

4. Press [STAT] and arrow over to the CALC menu.

Notice that the TI-83/84 Plus has two different linear regressions **4:LinReg(ax + b)** and **8:LinReg(a + bx)**. Number 8 most closely matches the statistics form so we will use it in this example. However, both are acceptable.

5. Press [8] or arrow down to **8:LinReg(a + bx)** and press [ENTER]. LinReg (a+bx) will appear at the top of your home screen.

6. The format for this command is LinReg(a+bx) xlist, ylist, Y-vars where the regression equation should be stored. Here we are using Lists 1 and 2 so press [2nd] L1 [,] [2nd] L2 for the lists. To get the Y where the equation will be stored, press [,] [VARS] arrow to Y-vars, press [ENTER] and then [ENTER] again to choose Y_1. Press [ENTER]. Your screen will appear as in Figure 14.26.

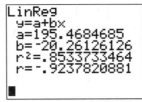

Figure 14.26

The value of r^2 is 0.853 when rounded to 3 decimal places. This indicates that a substantial amount of the variation in the observed prices is explained by regression, and therefore, that age is quite useful for predicting price.

14.3 Practice Problems (continued)

Problem 14.89 (continued)   Compute the coefficient of determination, r^2, for the age and price of Corvettes data given in Table 14.2.

Problem 14.91 (continued)   Compute the coefficient of determination, r^2, for the plant weight and quantity of volatile compounds emitted for potato plants data given in Table 14.3.

Problem 14.101 (continued)   Compute the coefficient of determination, r^2, for the per capita income and per capita beer consumption data provided on the WeissStats CD.

Problem 14.105 (continued)   Compute the coefficient of determination, r^2, for the estriol levels of pregnant women and birth weights data provided on the WeissStats CD.

Review
Problem #15 (continued)   Compute the coefficient of determination, r^2, for the student-to-faculty ratio and graduation rate data in Table 14.5

LESSON 14.4   LINEAR CORRELATION

Several statistics can be employed to measure the correlation between two variables. The one most commonly used is the **linear correlation coefficient, r,** also referred to as the **Pearson product moment correlation coefficient**. The linear correlation coefficient is a descriptive measure of the strength of the linear relationship between two variables.

Definition 14.7:   The **linear correlation coefficient, r,** of n data points is defined by

$$r = \frac{\frac{1}{n}\sum(x-\bar{x})(y-\bar{y})}{s_x s_y}$$

It can also be obtained from the computing formula

$$r = \frac{S_{xy}}{\sqrt{S_{xx}S_{yy}}}$$

where S_{xx}, S_{xy}, S_{yy} are given in Definition 14.3.

The TI-83/84 Plus can be used to compute the linear correlation coefficient without computing S_{xx}, S_{xy}, or S_{yy}.

Example 14.10   Compute the linear correlation coefficient, r, for the age and price of Orions data given in Table 14.1.

Solution: The default calculator setting is to not show r^2 or r when linear regression is done. Therefore we must first set our calculator to show it.

    1. Press 2nd followed by 0 to access the **CATALOG**. See Figure 14.27.

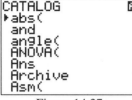 

Figure 14.27   Figure 14.28

    2. Press x⁻¹ (for the d-listings), arrow down to **DiagnosticOn** and press ENTER. Your calculator will return to the home screen with DiagnosticOn displayed.

    3. Press ENTER and the calculator will display the word Done. See Figure 14.28.

We will assume the ages are in List 1 and the prices are in List 2.

    4. Press STAT and arrow over to the CALC menu.

Notice that the TI-83/84 Plus has two different linear regressions **4:LinReg(ax + b)** and **8:LinReg(a + bx)**. Number 8 most closely matches the statistics form so we will use it in this example. However, both are acceptable.

    5. Press 8 or arrow down to **8:LinReg(a + bx)** and press ENTER. **LinReg (a+bx)** will appear at the top of your home screen.

    6. The format for this command is LinReg(a+bx) xlist, ylist, Y-vars where the regression equation should be stored. Here we are using Lists 1 and 2 so press 2nd L1 , 2nd L2 for the lists. To get the Y where the equation will be stored, press , VARS arrow to Y-vars, press ENTER and then ENTER again to choose Y_1. Press ENTER. Your screen will appear as in Figure 14.29.

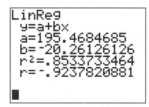

Figure 14.29

The value of r is -0.924 when rounded to 3 decimal places. This indicates there is a strong negative linear correlation between the age and price of Orions. In particular, it indicates that as age increases there is a strong tendency for price to decrease.

14.4 Practice Problems

Problem 14.123   Compute the linear correlation coefficient, r, for the age and price data of Corvettes given in Table 14.2.

Problem 14.125   Compute the linear correlation coefficient, r, for the plant weight and quantity of volatile compounds emitted for potato plants data in Table 14.3.

Problem 14.139   Compute the linear correlation coefficient, r, for the per capita income and per capita beer consumption data provided on the WeissStats CD.

Problem 14.139   Compute the linear correlation coefficient, r, for the estriol levels of pregnant women and birth weights data provided on the WeissStats CD.

Review   Compute the linear correlation coefficient, r, for the student-to-faculty ratio and graduation rate
Problem #16   data in Table 14.5

CHAPTER 15
INFERENTIAL METHODS IN REGRESSION AND CORRELATION

LESSON 15.1   THE REGRESSION MODEL; ANALYSIS OF RESIDUALS

To perform statistical inferences in regression and correlation, the variables under consideration must satisfy certain conditions. In this lesson, we will discuss those conditions and examine methods for checking whether they hold.

KEY FACT 15.1 ASSUMPTIONS (CONDITIONS) FOR REGRESSION INFERENCES:

1. **Population regression line:** There are constants β_0 and β_1 such that for each value x of the predictor variable, the conditional mean of the response variable is $\beta_0 + \beta_1 x$. We refer to the straight line $y = \beta_0 + \beta_1 x$ as the **population regression line** and to its equation as the **population regression equation**.

2. **Equal standard deviations:** The conditional standard deviations of the response variable are the same for all values of the predictor variable. We denote this common standard deviation by σ.

3. **Normal populations:** For each value of the predictor variable, the conditional distribution of the response variable is a normal distribution.

4. **Independent observations:** The observations of the response variable are independent of one another.

The inferential procedures in regression are robust to moderate violations of Assumptions 1-3 for regression inferences. In other words, the inferential procedures will work reasonably well provided the variables under consideration don't violate any of those assumptions too badly.

The sample regression line is used to estimate the population regression line. Of course, a sample regression line ordinarily will not be the same as the population regression line.

The statistic used to obtain a point estimate for the common conditional standard deviation σ is called the **standard error of the estimate** or the **residual standard deviation** and is defined as follows:

Definition 15.1: Standard Error of the Estimate

The standard error of the estimate s_e is defined by

$$s_e = \sqrt{\frac{SSE}{n-2}},$$

Where $SSE = \sum (y_i - \hat{y}_i)^2 = \sum y_i^2 - (\sum y_i)^2 / n$.

Very roughly speaking, the standard error of the estimate indicates how much, on average, the predicted values of the response variable differ from the observed values of the response variable.

Computing the standard error of the estimate is easy if SSE is known. If it is not known, but the sample regression equation is known, the TI-83/84 Plus may be used to compute it.

Example 15.2   The age and price data for a sample of 11 Orions is given in Table 15.1. Ages are in years, and prices are in hundreds of dollars, rounded to the nearest hundred dollars. The regression equation we found for this data is $\hat{y} = 195.47 - 20.26x$. Compute and interpret the standard error of the estimate, s_e.

Table 15.1	Car	1	2	3	4	5	6	7	8	9	10	11
	Age	5	4	6	5	5	5	6	6	2	7	7
	Price	85	103	70	82	89	98	66	95	169	70	48

Solution: We will assume that the ages are in List 1 and the prices are in List 2. Our first step is to compute SSE. The difference between an observed value and the predicted value is called a residual. The TI-83/84 Plus computes these residuals and stores them in a list called RESID. Therefore, to compute SSE we must square the values in this list and then sum them.

1. Execute the LinReg command. Your residuals are now stored in the list named RESID.

2. To sum the squares, press [2nd] LIST arrow over to MATH and press [5] for **sum(**.

3. Press [2nd] LIST and arrow to the list name RESID and press [ENTER].

4. Press [x²] [)] [ENTER]. This will square the list values and then sum them. See Figure 15.1.

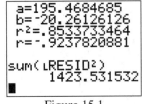
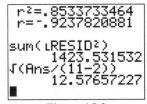

Figure 15.1   Figure 15.2

5. Find s_e by pressing [2nd] [x²] [2nd] ANS [÷] [(] 11 [−] 2 [)] [)] [ENTER]. Your screen should appear as in Figure 15.2. Note that we used the unrounded version of SSE so your answer may differ slightly from *Introductory Statistics* and *Elementary Statistics* on some problems.

15.1 Practice Problems

Problem 15.23   The *Kelley Blue Book* provides information on wholesale and retail prices of cars. Table 15.2 provides age and price data for 10 randomly selected Corvettes between 1 and 6 years old. Ages are given in years and prices are given in hundreds of dollars. Compute and interpret the standard error of the estimate, s_e.

Table 15.2	Age	6	6	6	2	2	5	4	5	1	4
	Price	270	260	274	405	364	295	335	308	405	305

Problem 15.25   Plants emit gases that trigger the ripening of fruit, attract pollinators, and cue other physiological responses. N. G. Agelopolous, K. Chamberlein, and J. A. Pickett examined factors that affect the emission of volatile compounds by the potato plant *Solanum tubersom* and published their findings in the *Journal of Chemical Ecology* (Vol. 26(2), pp. 497-511). The volatile compounds analyzed were hydrocarbons that are used by other plants and animals. Table 15.3 contains data on plant weight (x) in grams, and quantity of volatile compounds emitted (y), in hundreds of nanograms, for 11 potato plants. Compute and interpret the standard error of the estimate, s_e.

Table 15.3	x	57	85	57	65	52	67	62	80	77	53	68
	y	8.0	22.0	10.5	22.5	12.0	11.5	7.5	13.0	16.5	21.0	12.0

Problem 15.37   The magazine *Consumer Reports* publishes information on automobile gas mileage and variables that affect gas mileage. In the April 1999 issue, data on gas mileage (in mpg) and engine displacement (in liters, L) were published for 121 vehicles. Those data are stored on the WeissStats CD. Compute and interpret the standard error of the estimate, s_e.

Problem 15.38   J. Greene and J. Touchstone conducted a study on the relationship between the estriol levels of pregnant women and the birth weights of their children. Their findings, "Urinary Tract Estriol: An Index of Placental Function," were published in the *American Journal of Obstetrics and Gynecology* (Vol. 85(1), pp. 1-9). The data from the study are provided on the WeissStats CD, where estriol levels are in mg/24 hr and birth weights are in hectograms. Compute and interpret the standard error of the estimate, s_e.

Review Problem # 12   Graduation rate- the percentage of entering freshmen, attending full time, that graduate within 5 years- and what influences it have become a concern in U.S. colleges and universities. *U.S. News and World Report's* "College Guide" provides data on graduation rates for colleges and universities as a function of the percentage of freshmen in the top 10% of their high-school class, total spending per student, and student-to-faculty ratio. A random sample of 10 universities gave the data in Table 15.4 on student-to-faculty ratio (S/F ratio) and graduation rate (grad rate). Compute and interpret the standard error of the estimate, s_e.

Table 15.4

S/F ratio x	Grad rate y
16	45
20	55
17	70
19	50
22	47
17	46
17	50
17	66
10	26
18	60

The method for checking Assumptions 1-3 relies on an analysis of the errors made in using the regression equation to predict the observed values of the response variable. Each such difference is called a residual, denoted generically by the letter *e*. Recall that the TI-83/84 Plus stores these values in a list called RESID every time the LinReg command is used.

KEY FACT 15.2 RESIDUAL ANALYSIS FOR THE REGRESSION MODEL:

If the assumptions for regression inferences are met, then the following two conditions should hold:

- A plot of the residuals, called a **residual plot**, against the values of the predictor variable should fall roughly in a horizontal band centered and symmetric about the x-axis.

- A normal probability plot of the residuals should be roughly linear.

Failure of either of these two conditions casts doubt on the validity of one or more of the assumptions for regression inferences for the variables under consideration.

Example 15.3   Perform a residual analysis to decide whether it is reasonable to consider the assumptions for regression inferences met by the variables age and price of Orions given in Table 15.1.

Solution: We assume that age is in List 1 and that the residuals are stored in the list RESID. (To accomplish this, run a linear regression on the age and price and the residuals will be stored there.)

First we will plot the residuals against age. We will use the scatterplot to do this. Be sure to clear the Y= screen or turn off any equations located there. Also, be sure all stat plots are turned off except Plot 1.

1. Press 2nd STATPLOT ENTER to enter the Stat Plot setup screen.

2. Highlight On and press ENTER.

3. Select the scatterplot by highlighting the icon ⊾∵ and pressing ENTER.

4. At the Xlist prompt press 2nd L1 for List 1 (the ages).

5. At the Ylist prompt press 2nd LIST, arrow down to RESID and press ENTER.

6. Choose your mark and press ENTER. Here the box is chosen. Your screen should appear as in Figure 15.3.

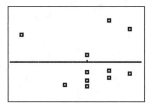

Figure 15.3   Figure 15.4

7. Use ZoomStat to set the window by pressing ZOOM 9. Your screen should appear as in Figure 15.4.

Taking into account the small sample size, we can say that the residuals fall roughly in a horizontal band centered and symmetric about the x-axis.

Now we must do the normal probability plot of the residuals.

8. Press 2nd STATPLOT ENTER to enter the Stat Plot setup screen.

9. Highlight On and press ENTER.

10. Select the normal plot by highlighting the icon ∠ and pressing ENTER.

11. For Data List we want to use RESID so press 2nd LIST, arrow down to RESID, and press ENTER.

12. For Data Axis, highlight X and press ENTER.

13. Choose your mark and press ENTER. Your screen should appear as in Figure 15.5.

14. Use ZoomStat to set the window by pressing ZOOM 9. Your screen should appear as in Figure 15.6. Note with the Axes turned off it will appear as in Figure 15.7.

Figure 15.5

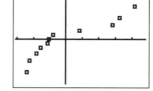
Figure 15.6

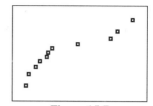
Figure 15.7

The normal probability plot of the residuals is very roughly linear, although the departure from linearity is sufficient for some concern. However, the inferential procedures in regression are robust to moderate violations of the assumptions for regression inferences, so we will consider the assumptions met for this data.

15.1 Practice Problems (continued)

Problem 15.23 (continued)   Perform a residual analysis to decide whether it is reasonable to consider the assumptions for regression inferences met by the variables age and price of Corvettes given in Table 15.2.

Problem 15.25 (continued)   Perform a residual analysis to decide whether it is reasonable to consider the assumptions for regression inferences met by the variables plant weight and quantity of volatile compounds emitted for potato plants data in Table 15.3.

Problem 15.37 (continued)   Perform a residual analysis to decide whether it is reasonable to consider the assumptions for regression inferences met by the variables gas mileage and engine displacement provided on the WeissStats CD.

Problem 15.38 (continued)   Perform a residual analysis to decide whether it is reasonable to consider the assumptions for regression inferences met by the variables estriol levels of pregnant women and birth weight of children provided on the WeissStats CD.

Review
Problem #13
Perform a residual analysis to decide whether it is reasonable to consider the assumptions for regression inferences met by the variables student-to-faculty ratio and graduation rate given in Table 15.4.

LESSON 15.2 INFERENCES FOR THE SLOPE OF THE POPULATION REGRESSION LINE

The inferential techniques we will discuss in this section require that the assumptions for regression inferences given in Key Fact 15.1 are satisfied by the variables under consideration. However, as we noted earlier, these techniques are robust to moderate violations of those assumptions.

Hypothesis Tests for the Slope of the Population Regression Line

Of particular interest is whether the slope, β_1, of the population regression line equals 0. Because if $\beta_1 = 0$, then for each value x of the predictor variable, the conditional distribution of the response variable is a normal distribution having mean β_0 and standard deviation σ; and neither of those two parameters involves x. Consequently, in this case, x is useless as a predictor of y.

Example 15.5   The data on age and price for a sample of 11 Orions are displayed in Table 15.1. At the 5% significance level, do the data provide sufficient evidence to conclude that age is useful as a predictor of price for Orions?

Solution: As we discovered in Example 15.2, it is reasonable to consider the assumptions for regression inferences satisfied by the variables age and price for Orions, at least for Orions between 2 and 7 years old.

Let β_1 denote the slope of the population regression line that relates price to age for Orions.

Step 1: State the null and alternative hypotheses

H_0: $\beta_1 = 0$ (age is not useful for predicting price)

H_a: $\beta_1 \neq 0$ (age is useful for predicting price)

Step 2: Decide on the significance level.

We are to perform the hypothesis test at the 5% significance level; so $\alpha = 0.05$.

Step 3: Compute the value of the test statistic

We will assume that the ages are in List 1 and the prices in List 2.

1. Press [STAT], arrow over to TESTS, and press [ALPHA] E or arrow down to **E:LinRegTTest** and press [ENTER].

2. Enter Xlist as List 1, Ylist as List 2, and Freq as 1.

3. Highlight the correct alternative hypothesis and press [ENTER]. Here we wish to highlight $\neq$.

4. The RegEQ: line allows you to store the regression equation to a particular Y variable. We will store it to Y_1 by pressing [VARS], arrowing to Y-VARS, pressing [ENTER] and then [ENTER] again. Your screen should appear as in Figure 15.8.

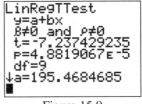

Figure 15.8   Figure 15.9

5. Highlight Calculate and press [ENTER]. Your screen will appear as in Figure 15.9.

Our t-value is given as -7.237 when rounded to 3 decimal places

Step 4: Obtain the *p*-value.

The *p*-value is approximately 0.00005. Note because the calculator uses unrounded values for b_1, s_e, and S_{xx}, these values may differ slightly from the textbook.

Step 5: If *p*-value < α reject H_0; otherwise do not reject H_0.

Since *p*-value is less than the significance level, we reject H_0.

Step 6: Interpret the results of the hypothesis test.

At the 5% significance level, the data provide sufficient evidence to conclude that the slope of the population regression line is not 0 and hence that age is useful as a predictor price for Orions.

15.2 Practice Problems

Problem 15.51   The age and price of Corvettes are displayed in Table 15.2. At the 5% significance level, do the data provide sufficient evidence to conclude that age is useful as a predictor of price for Corvettes?

Problem 15.53   The plant weight and quantity of volatile compounds emitted for potato plants is displayed in Table 15.3. At the 5% significance level, do the data provide sufficient evidence to conclude that weight is useful as a predictor of quantity of volatile compounds emitted for potato plants?

Problem 15.69   The gas mileage and engine displacement data are provided on the WeissStats CD. At the 5% significance level, do the data provide sufficient evidence to conclude engine displacement is a useful predictor of gas mileage?

Problem 15.70   The estriol levels of pregnant women and birth weights of their children data are provided on the WeissStats CD. At the 5% significance level, do the data provide sufficient evidence to conclude that estriol levels of pregnant women is useful as a predictor of birth weight of children?

Problem   The student-to-faculty ratio and graduation rate data are displayed in
Problem #14   Table 15.4. At the 5% significance level, do the data provide sufficient evidence to conclude that student-to-faculty ratio is useful as a predictor of graduation rate?

Confidence Intervals for the Slope of the Population Regression Line

It is worthwhile to obtain an estimate for the slope of the population regression line. The slope represents the change in the conditional mean response variable for each increase in the value of the predictor variable. In our example of the Orions, the slope is the amount that the mean price decreases for every increase in age by 1 year. In other words, the slope is the mean yearly depreciation of the Orions.

The confidence interval for the slope of the population regression line assumes the assumptions for regression inference are met.

Procedure 15.2: For a confidence level of $1 - \alpha$, the endpoints of the confidence interval are

$$b_1 \pm t_{\alpha/2} \cdot \frac{s_e}{\sqrt{S_{xx}}}$$

with df = n - 2.

Example 15.6   Use the data in Table 15.1 to obtain a 95% confidence interval for the slope of the population regression line that relates price to age for Orions.

Solution: The TI-84 Plus has a built in procedure for this interval. The TI-83 Plus does not. However, it can be used to help compute it. We will demonstrate both beginning with the TI-84.

Using the TI-84 Plus

We assume the x-values are in List 1 and the y-values are in List 2.

1. Press [STAT], arrow over to TESTS, and press [ALPHA] G or arrow down to **G:LinRegTTest** and press [ENTER].

2. Enter Xlist as List 1, Ylist as List2, and Freq as 1.

3. Enter your confidence level of 0.95.

4. The RegEQ line allows you to store the regression equation to a particular Y variable. We will sore it to Y_1 by pressing VARS, arrow over to Y-VARS, press ENTER and then press ENTER again. Your screen should appear as in Figure 15.10.

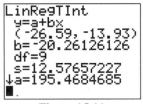

Figure 15.10   Figure 15.11

5. Highlight Calculate and press ENTER. Your screen will appear as in Figure 15.11.

Using the TI-83 Plus

To begin, we must find the values to plug into our formula.

1. To find the b_1 value, we can run the LinRegTTest. This will also give us the value of s_e. The instructions for running this test are given in Example 15.3. The resulting screen is shown in Figure 15.12. Note you will have to arrow down to see this information.

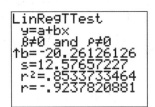

Figure 15.12

Based on this screen, b_1 is -20.26 and s_e = 12.58 when rounded to two decimal places.

2. Because our n = 11 our df = 9. We can find the t-value by running the program INVT. Press PRGM arrow down to INVT and press ENTER. prgmINVT will appear on your home screen.

3. Press ENTER to run the program.

4. Follow the prompts of the command entering 9 for DF, and 2 for C because we want the area to the right. For the area, remember that we always want to enter the area to the right which is half of α. Therefore for the area enter 0.05/2 = 0.025. The results displayed will be your t-value. See Figure 15.13.

Figure 15.13

In this case our t-value is 2.262 when rounded to three decimal places.

To compute S_{xx}, the formula is

$$S_{xx} = \sum x^2 - \left(\sum x\right)^2 / n$$

Therefore, we will need the sum of the x-values squared as well as the square of the sum of the x-values. These can be obtained by running 1-Var Stats on List 1.

5. Press [STAT] arrow over to CALC and press [ENTER].

6. Enter the name of the list where the x-values are stored. In this case, we are using List 1 so press [2nd] L1 [ENTER]. The sums we are interested in will be displayed. See Figure 15.14.

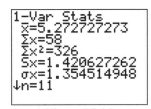

Figure 15.14

7. Compute S_{xx} by entering $326 - (58)^2 / 11$ and pressing [ENTER]. The result of 20.18 will be our S_{xx}.

8. Compute the left endpoint of the confidence interval by entering

$$-20.26 - 2.262 * 12.58 / \sqrt{(20.18)} \text{ and pressing [ENTER].}$$

9. Compute the right endpoint of the confidence interval by pressing [2nd] ENTRY, arrowing to the - sign and changing it to +. Then press [ENTER]. The results are shown in Figure 15.15.

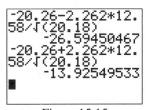

Figure 15.15

The interval is -26.59 to -13.93. We can be 95% confident that the slope, β_1, of the population regression line is somewhere between -26.59 and -13.93. In other words, we can be 95% confident that the yearly decrease in mean price for Orions is somewhere between $1393 and $2659.

15.2 Practice Problems (continued)

Problem 15.57   Use the data in Table 15.2 to obtain a 99% confidence interval for the slope of the population regression line that relates age and price of Corvettes.

Problem 15.59   Use the data in Table 15.3 to obtain a 95% confidence interval for the slope of the population regression line that relates plant weight and quantity of volatile compounds emitted for potato plants.

Problem 15.21   Use the data in Table 15.4 to obtain a 90% confidence interval for the slope of the population regression line that relates age and percent body fat.

Problem 15.22   Use the data provided on the WeissStats CD to obtain a 99% confidence interval for the slope of the population regression line that relates gas mileage and engine displacement.

Review Problem # 14 (continued)   Use the data in Table 15.5 to obtain a 90% confidence interval for the slope of the population regression line that relates student-to-faculty ratio and graduation rate.

LESSON 15.3 ESTIMATION AND PREDICTION

In regression analysis there are two important types of inferences:

1.   to estimate the conditional mean of the response variable corresponding to a particular value of the predictor variable and

2.   to predict the value of the response variable for an individual value of the predictor variable.

In the first case, we use a **confidence interval for the mean response** at a particular value of the predictor. For example, in the Orion data, we could obtain a confidence interval for the mean price of all 3-year-old Orions. The confidence interval is estimating a *mean* value.

In the second case, we use a **prediction interval for an individual response** at a particular value of the predictor. For example, in the Orion data, we could obtain a prediction interval for the price of a 3-year-old Orion. The prediction interval is estimating an *individual* value.

A **prediction interval** is slightly wider than a **confidence interval**. It makes sense that the price of a single Orion would be more variable than the price of the mean of all Orions of the same age.

LESSON 15.4 INFERENCES IN CORRELATION

Frequently we want to decide whether two variables are linearly correlated, that is, whether there is a linear relationship between the two variables. We can perform a hypothesis test for the **population linear correlation coefficient, ρ**. The population linear correlation coefficient measures the linear correlation of all possible pairs of observations of two variables. If $\rho > 0$, the variables are **positively linearly correlated** and if $\rho < 0$, the variables are negatively linearly correlated. If $\rho = 0$, the variables are **linearly uncorrelated**.

The TI-83/84 Plus has a built in procedure for testing the population linear correlation coefficient.

Example 15.11   Refer to the age and price data for a sample of 11 Orions given in Table 15.1. At the 5% significance level, do the data provide sufficient evidence to conclude that age and price of Orions are negatively linearly correlated?

Solution: As we discovered in Example 15.3, it is reasonable to consider the assumptions for regression inferences satisfied by the variables age and price for Orions, at least for Orions between 2 and 7 years old.

Let ρ denote the population linear correlation coefficient for the variables age and price of Orions. Then the null and alternative hypotheses are:

Step 1: State the null and alternative hypotheses

H_0: $\rho = 0$ (age and price are linearly uncorrelated)

H_a: $\rho \neq 0$ (age and price are negatively linearly correlated.)

Step 2: Decide on the significance level.

We are to perform the hypothesis test at the 5% significance level; so $\alpha = 0.05$.

Step 3: Compute the value of the test statistic.

We will assume that the ages are in List 1 and the prices in List 2.

1. Press [STAT], arrow over to TESTS and press [ALPHA] E or arrow down to **E:LinRegTTest** and press [ENTER].

2. Enter Xlist as List 1, Ylist as List 2, and Freq as 1.

3. Highlight the correct alternative hypothesis and press [ENTER]. Here we wish to highlight < 0.

4. The RegEQ: line allows you to store the regression equation to a particular Y variable. We will store it to Y_1 by pressing [VARS] arrowing to Y-VARS, pressing [ENTER] and then [ENTER] again. Your screen should appear as in Figure 15.16.

Figure 15.16

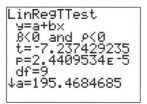
Figure 15.17

5. Highlight Calculate and press [ENTER]. Your screen will appear as in Figure 15.17.

Our t-value is given as -7.237 when rounded to 3 decimal places

Step 4: Obtain the *p*-value.

The *p*-value is approximately 0.00002. Note because the calculator uses unrounded values for b_1, s_e, and S_{xx}, these values may differ slightly from the textbook.

Step 5: If *p*-value $\leq \alpha$, reject H_0; otherwise do not reject H_0.

Since *p*-value is less than the significance level, we reject H_0.

Step 6: Interpret the results of the hypothesis test.

At the 5% significance level, the data provide sufficient evidence to conclude that age and price of Orions are negatively linearly correlated. Prices for Orions tend to decrease linearly with increasing age, at least for Orions between 2 and 7 years old.

15.4 Practice Problems

Problem 15.109   Refer to the age and price of Corvettes data given in Table 15.2. At the 10% significance level, do the data provide sufficient evidence to conclude that age and price are negatively linearly correlated for Corvettes?

Problem 15.111   Refer to the plant weight and quantity of volatile compounds emitted for potato plants data given in Table 15.3. At the 5% significance level, do the data provide sufficient evidence to conclude that plant weight and quantity of volatile compounds emitted for potato plants are positively linearly correlated?

Problem 15.123   Refer to the gas mileage and engine displacement data provided on the WeissStats CD. At the 1% significance level, do the data provide sufficient evidence to conclude that gas mileage and engine displacement are linearly correlated?

Problem 15.124   Refer to the estriol levels of pregnant women and birth weights of children data provided on the WeissStats CD. At the 10% significance level, do the data provide sufficient evidence to conclude that estriol levels of pregnant women and birth weights of children are linearly correlated?

Review   Refer to the student-to-faculty ratio and graduation rate data given in
Problem #16   Table 15.4. At the 5% significance level, do the data provide sufficient evidence to conclude that student-to-faculty ratio and graduation rate data are positively linearly correlated?

LESSON 15.5   TESTING FOR NORMALITY*

As we know, several descriptive methods are available for assessing normality of a variable from sample data. One of the most commonly used methods is the normal probability plot. If the variable is normally distributed, then a normal probability plot of the sample data should be roughly linear. This visual assessment of normality is subjective because what constitutes "roughly linear" is a matter of opinion. To overcome this difficulty we can perform a hypothesis test for normality based on the linear correlation coefficient. Procedure 15.6 outlines the steps to perform a Correlation Test for Normality. The assumption is a simple random sample.

Example 15.13   The Internal Revenue Service publishes data on federal individual income tax returns in *Statistics of Income, Individual Income Tax Returns*. A random sample of 12 returns from last year revealed the adjusted gross incomes, in thousands of dollars, shown in Table 15.5. At the 5% significance level, do the data provide sufficient evidence to conclude that adjusted gross incomes are not normally distributed?

Table 15.5	9.7	93.1	33.0	21.2
	81.4	51.1	43.5	10.6
	12.8	7.8	18.1	12.7

Solution:

Step 1: State the null and alternative hypotheses
H_0: Adjusted gross incomes are normally distributed.
H_a: Adjusted gross incomes are not normally distributed.

Step 2: Decide on the significance level.
We are to perform the hypothesis test at the 5% significance level; so $\alpha = 0.05$.

Step 3: Compute the value of the test statistic
Our first step is to enter our data, with the corresponding normal scores into the calculator. Because the data must be in ascending order, along with the normal scores, we will enter the incomes and then sort them.

1. With the incomes data entered in List 1, press STAT and press 2 or arrow down to **2:SortA(** and press ENTER. The command will appear on your home screen.

2. Enter the list you wish to sort, in this case List 1. Press 2nd L1) ENTER. The calculator will display Done to indicate the data has been sorted.

3. Now we must enter the normal scores into List 2. Find these in Table III in Appendix A of the textbook. These scores are listed below in Table 15.6.

4. Do a linear regression on the data values and their normal scores pressing STAT arrowing over to CALC, pressing 8 for LinReg(a + bx), 2nd L1 , 2nd L2 for Lists 1 and 2, and pressing ENTER. The r-value that is displayed is our test statistic.

The value of the test statistic is $R_p = 0.908$.

Step 4: Obtain the Critical Value.
To find the critical value, consult Table XIV in *Introductory Statistics* and *Elementary Statistics*. For this problem, the critical value is 0.927. We will reject if $R_p < 0.927$.

Step 5: Compare test statistic to critical value.
The value of the test statistic is $R_p = 0.908$ which is less than the critical value of 0.927. Therefore we reject H_0.

Step 6: Interpret the results of the hypothesis test.
At the 5% significance level, the data provide sufficient evidence to conclude that adjusted gross incomes are not normally distributed.

Table 15.6

Adjusted gross income	Normal score
7.8	-1.64
9.7	-1.11
10.6	-0.79
12.7	-0.53
12.8	-0.31
18.1	-0.10
21.1	0.10
33.0	0.31
43.5	0.53
51.1	0.79
81.4	1.11
93.1	1.64

15.5 Practice Problems

Problem 15.131   A sample of the final exam scores in a large introductory statistics course is displayed in Table 15.7. At the 5% significance level, do the data provide sufficient evidence to conclude the final exam scores are not normally distributed?

Table 15.7

88	67	64	76	86
85	82	39	75	34
90	63	89	90	84
81	96	100	70	96

Problem 15.13   Table 15.8 displays finishing times, in seconds, for the winners of fourteen 1-mile thoroughbred horse races, as found in two recent issues of *Thoroughbred Times*. At the 1% significance level do the data provide sufficient evidence to conclude that the finishing times for the winners of 1-mile thoroughbred horse races are not normally distributed?

Table 15.8

94.15	93.37	103.02	95.57	97.73	101.09	99.38
97.19	96.63	101.05	97.91	98.44	97.47	95.10

CHAPTER 16
ANALYSIS OF VARIANCE (ANOVA)

LESSON 16.1   THE F-DISTRIBUTION

A variable is said to have an F-distribution if its distribution has the shape of a special type of right-skewed curve, called an F-curve. There are infinitely many F-distributions, and we identify the F-distribution in question by stating its degrees of freedom. An F-distribution has two numbers of degrees of freedom instead of one. The first number of degrees of freedom for an F-curve is called the degrees of freedom for the numerator and the second the degrees of freedom for the denominator. An F-distribution starts at 0 and goes infinitely to the right.

The TI-83/84 Plus has a built in function to find the area under a F-curve with specified degrees of freedom to the left of a F-value.

Example   For a F-curve with df = (8, 5), find the area under the curve to the left of 4.82.

Solution:

> 1. Press $\boxed{\text{2nd}}$ DISTR to access the distribution menu.

> 2. Press $\boxed{9}$ or arrow down to **9:Fcdf(** and press $\boxed{\text{ENTER}}$.

> 3. The format of this command is **Fcdf**(lowerbound, upperbound, num, den). For our problem we need to find the area from 0 to 4.82. For our problem, our command is therefore **Fcdf**(0, 4.82, 8, 5). The result is shown in Figure 16.1.

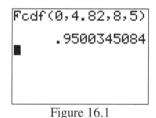

Figure 16.1

If you are looking for the area to the right of a F-value, you will need to estimate ∞ as 1E99. Keep in mind that unlike the normal and t-curves, the F-curve starts at 0 and is right skewed, although the total area under the curve is equal to 1.

The TI-83/84 Plus does not have a built-in InvF function like InvNorm. However, the program INVF contained in the WeissStats CD can be used to accomplish the same thing.

Example 16.1   For an F-curve with df = (4, 12), find $F_{0.05}$; that is the F-value having area 0.05 to its right.

Solution:

> 1. Press $\boxed{\text{PRGM}}$ and arrow down to INVF and press $\boxed{\text{ENTER}}$. prgmINVF should appear on your home screen.

> 2. Press $\boxed{\text{ENTER}}$ to run the program.

> 3. Follow the prompts entering 4 for the df of the numerator, 12 for the df of the denominator, 2 for area to the right, and 0.05 for the area. The calculator will take a little while to do the computation. See Figure 16.2 for the final result.

Figure 16.2

For an F-curve with df = (4, 12), the F-value having area 0.05 to the right is 3.26 when rounded to two decimal places.

16.1 Practice Problems

Problem 16.7 For an F-curve with df = (24, 30), find the F-value having the specified area to its right.
　　　a) 0.05　　　　b) 0.01　　　　c) 0.025

Problem 16.8 For an F-curve with df = (12, 5), find the F-value having the specified area to its right.
　　　a) 0.01　　　　b) 0.05　　　　c) 0.005

Problem 16.9 For an F-curve with df = (20, 21), find
　　　a) $F_{0.01}$　　　　b) $F_{0.05}$　　　　c) $F_{0.10}$

Problem 16.10 For an F-curve with df = (6, 10), find
　　　a) $F_{0.05}$　　　　b) $F_{0.01}$　　　　c) $F_{0.025}$

LESSON 16.2 ONE-WAY ANOVA: THE LOGIC

Analysis of Variance (ANOVA) provides methods for comparing several population means--that is, the means of a single variable for several populations. First, we will study the simplest kind of ANOVA, one-way analysis of variance. One-way analysis of variance is the generalization to more than two populations of the pooled t-procedure. Procedure 16.1 outlines the steps for performing a one-way ANOVA. As in the pooled t-procedure, we must make the following assumptions:

　　1. Simple random samples.

　　2. Independent samples: The samples taken from the populations under consideration are independent of one another.

　　3. Normal populations: For each population, the variable under consideration is normally distributed.

　　4. Equal standard deviations: The standard deviations of the variable under consideration are the same for all the populations.

Generally, normal probability plots are effective in detecting gross violations of the normality assumption. The equal-standard-deviations assumption is usually more difficult to check. As a rule of thumb, we consider this assumption satisfied if *the ratio of the largest to the smallest standard deviation is less than 2*.

LESSON 16.3 ONE-WAY ANOVA: THE PROCEDURE

Example 16.3 Independent simple random samples of households in the four U.S. regions yielded the data on last year's energy consumptions shown in Table 16.1. At the 5% significance level, do the data provide sufficient evidence to conclude that a difference exists in last year's mean energy consumption by households among the four U.S. regions?

Table 16.1

Northeast	Midwest	South	West
15	17	11	10
10	12	7	12
13	18	9	8
14	13	13	7
13	15		9
	12		

Solution: First we check that the four conditions required for performing a one-way ANOVA are satisfied. The samples are independent. Normal probability plots (not shown) of the four samples reveal no outliers and are roughly linear. In addition, the rule of 2 is satisfied, so all the requirements are met.

Let μ_1, μ_2, μ_3, and μ_4 denote last year's mean energy consumptions for households in the Northeast, Midwest, South, and West, respectively. We want to perform the hypothesis test.

Step 1: State the null and alternative hypotheses.
   H_0: $\mu_1 = \mu_2 = \mu_3 = \mu_4$ (mean energy consumptions are equal)
   H_a: Not all the means are equal.

Step 2: Decide on the significance level, α .
The test is to be performed at the 5% significance level. Thus $\alpha = 0.05$.

Step 3: Compute the value of the test statistic.
To use the TI-83/84 PLUS, we must first store the samples in Table 16.1 in 4 lists.

1. With the samples for the Northeast, Midwest, South, and West stored in Lists 1-4, respectively, press STAT and arrow over to the TEST menu.

2. Press ALPHA F or arrow up to **F:ANOVA(** and press ENTER.

3. **ANOVA(** will appear on your home screen. Enter the names of the lists where your samples are stored separated by commas. For this example our command is
ANOVA(L_1, L_2, L_3, L_4).

4. Press ENTER. Your screen will appear as in Figure 16.3. Note that the format of the information is a little different from the standard table. You will have to arrow down the screen to see all the information. See Figure 16.4.

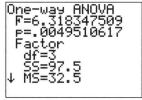
Figure 16.3

Figure 16.4

The test statistic is F= 6.318. It has an F-distribution with df = (3, 16).

Step 4: Obtain the *p*-value.
The first screen shows our *p*-value = 0.005 (to three decimal places).

Step 5: If *p*-value ≤ α , reject H_0; otherwise, do not reject H_0.
Because the *p*-value is less than the specified significance level of 0.05, we reject H_0.

Step 6: Interpret the results of the hypothesis test.
The data provide sufficient evidence to conclude that last year's mean energy consumptions for households in the four U.S. regions are not all the same.

16.3 Practice Problems

Problem 16.49   Copepods are tiny crustaceans that are an essential link in the estuarine food web. Marine scientists G. Weiss, G. McManus, and H. Harvey at the Chesapeake Biological Laboratory in Maryland designed an experiment to determine whether dietary lipid (fat) content is important in the population growth of a Chesapeake Bay copepod. Their findings were published as the paper "Development and Lipid Composition of the Harpacticoid Copepod Nitocra Spinipes Reared on Different Diets" (*Marine Ecology Progress Series*, Vol. 132, pp. 57-61). Independent random samples of copepods were placed in containers containing lipid-rich diatoms, bacteria, or leafy macroalgae. There were 12 containers total, four replicates per diet. Five gravid (egg-bearing) females were placed in each container. After 14 days, the number of copepods in each container were counted. These counts are given in Table 16.2. At the 5% significance level, do the data provide sufficient evidence to conclude that a difference exists in mean number of copepods among the three different diets?

Table 16.2	Diatoms	426	467	438	497
	Bacteria	303	301	293	328
	Macroalgae	277	324	302	272

Problem 16.70   Advertising researchers Shsuptrine and McVicker wanted to determine whether there were significant differences in the readability of magazine advertisements. Thirty magazines were classified based on their educational level- high, mid, or low- and then three magazines were randomly selected from each level. From each magazine, six advertisements were randomly chosen and examined for readability. In this particular case, readability was characterized by the numbers of words, sentences and three syllable or more words in each ad. The researchers published their findings in the *Journal of Advertising Research* (Vol. 21, p. 47). The number of three syllable or more words in each ad are provided on the WeissStats CD. At the 5% significance level, do the data provide evidence of a difference in the mean number of three syllable or more words among the three magazine levels?

Review   The U.S. Federal Bureau of Investigation conducts surveys to obtain information on the value of
Problem #19   losses due to various types of robberies. Results of the surveys are published in *Population-at-Risk Rates and Selected Crime Indicators*. Independent random samples of reports for three types of robberies- highway, gas station, and convenience store- gave the data in Table 16.3, in dollars, on value of losses. At the 5% significance level, do the data provide sufficient evidence to conclude that a difference in mean losses exists among the three types of robberies?

Table 16.3	Highway	Gas station	Convenience store
	952	1298	844
	996	1195	921
	839	1174	880
	1088	1113	706
	1024	953	602
		1280	614

LESSON 16.4   MULTIPLE COMPARISONS*

Suppose we perform a one-way ANOVA and reject the null hypothesis. Then we can conclude that the means of the populations under consideration are not all the same. We may want to know which means are different or the relation among all the means. Methods for dealing with these problems are called **multiple comparisons.** In *Introductory Statistics*, the **Tukey multiple-comparison** is discussed.

One approach for implementing multiple comparisons is to obtain confidence intervals. Two means are declared different if the confidence interval for their difference does not contain 0.

Procedure 16.2 of your textbook outlines the steps for performing a Tukey multiple-comparison. The assumptions are the same as for the one-way ANOVA, namely:

    1. Simple random samples

    2. Independent samples

    3. Normal populations

    4. Equal population standard deviations

Formula 16.1: The endpoints of the Tukey multiple-comparison confidence interval for $\mu_i - \mu_j$ are

$$\left(\overline{x}_i - \overline{x}_j\right) \pm \frac{q_\alpha}{\sqrt{2}} \cdot s \sqrt{\left(\frac{1}{n_i}\right) + \left(\frac{1}{n_j}\right)},$$

where $s = \sqrt{MSE}$. Do this for all possible pairs of means with $i < j$.

The program Tukey contained in the WeissStats CD can be used to compute the intervals. The Tukey multiple-comparison method is based upon the studentized range distribution or q-distribution. The TI-83/84 Plus does not have this distribution built-in, so you will need to obtain the values of q from the tables in *Introductory Statistics*.

Example 16.6   Energy Consumption: Apply the Tukey multiple-comparison method to the energy consumption data, repeated here in Table 16.4.

Table 16.4

Northeast	Midwest	South	West
15	17	11	10
10	12	7	12
13	18	9	8
14	13	13	7
13	15		9
	12		

Solution: We must first decide on the family confidence level. For this illustration we will use $\alpha = 0.05$; so the family confidence level is 0.95 (95%).

Using Table XIII in the appendix of *Introductory Statistics*, we find $q_{0.05}$ with parameters 4 and 16 to be 4.05.

To run the program Tukey, the sample data must all be in List 1 with the number corresponding to the sample located in List 2. For this example, we will let Northeast be sample 1, Midwest sample 2, South sample 3, and West sample 4. You may find it easier to enter all the Northeast data into List 1, and then enter the corresponding 1's in List 2 before entering the Midwest data in List 1. A portion of the list editor with the data entered is shown below in Figure 16.5.

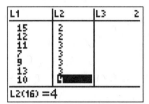

Figure 16.5

1. From the home screen, press PRGM, arrow down to the program Tukey, and press ENTER. prgmTUKEY will appear on your home screen.

2. Press ENTER to run the program.

3. Enter the q-value obtained from *Introductory Statistics* and press ENTER. Here the value is 4.05. The program will take some time to run. The final output is displayed in Figure 16.6.

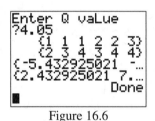

Figure 16.6

The first list is the first sample for the interval and the second list is the second sample. To see all the intervals it will be necessary to view the stat list editor. Lists 3 and 4 contain the sample numbers while List 5 contains the sample sizes for the samples in numerical order. List 6 contains the means of the samples in numerical order.

The intervals are contained in the lists MINUS and PLUS. MINUS contains the left endpoint of the interval and PLUS contains the right endpoint of the interval. To view these lists in the stat editor, use the following steps.

4. Press STAT ENTER to enter the list editor.

5. Arrow to where you can see Lists 3 and 4. These tell you the samples used to the compute the interval. List 3(I) is the first sample while List 4(I) is the second sample for the interval in Minus(I) and Plus(I). You may wish to write these down.

6. Arrow to the top of a list and highlight the list number. Then arrow beyond the last list to a blank list.

7. Enter the name of the list you want to view, here MINUS, by pressing 2nd LIST and arrow over to the name of the list and pressing ENTER twice. The list should appear in the edit screen.

8. Repeat steps 6 and 7, but use PLUS instead of MINUS. Figure 16.7 displays the final edit screen.

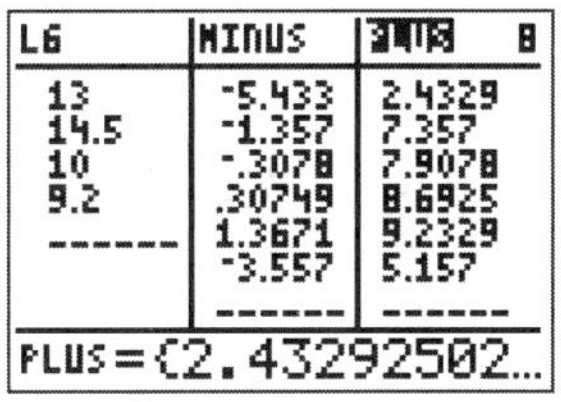

L6	MINUS	PLUS	8
13	-5.433	2.4329	
14.5	-1.357	7.357	
10	-.3078	7.9078	
9.2	.30749	8.6925	
-----	1.3671	9.2329	
	-3.557	5.157	
	------	------	
PLUS= {2.43292502...			

Figure 16.7

The intervals are now shown. For example, the first elements in Lists MINUS and PLUS correspond to the interval for the Northeast and Midwest samples. The interval is (-5.43, 2.43).

The 95% confidence intervals are shown simultaneously below in Table 16.5.

	Northeast (1)	Midwest (2)	South (3)
Midwest (2)	(-5.43, 2.43)		
South (3)	(-1.36, 7.36)	(0.31, 8.69)	
West (4)	(-0.31, 7.91)	(1.37, 9.23)	(-3.56, 5.16)

Table 16.5

By viewing List 6 we can see the means of the samples. The means in sample order are 13, 14.5, 10, and 9.2. We will rank the means from smallest to largest and connect by lines those whose populations means were not declared different. If the confidence interval does not contain 0, then declare the means different.

West (4)	South (3)	Northeast (1)	Midwest (2)
9.2	10.0	13.0	14.5

Interpreting this diagram, we conclude that last year's mean energy consumption in the Midwest exceeds that in the West and South, and that no other means can be declared different. *All of this* can be said with 95% confidence, the family confidence level.

16.4 Practice Problems

Problem 16.95   Refer to the data in Problem 16.49. Perform a Tukey multiple-comparison using a family confidence level of 0.95.

Problem 16.116   Refer to the data on the number of three syllable or more words in advertising in Problem 16.70. The data is contained on the WeissStats CD. Perform a Tukey multiple-comparison using a family confidence level of 0.95.

Review
Problem #23
Refer to the data in Review Problem #19. Perform a Tukey multiple-comparison using a family confidence level of 0.95.

LESSON 16.5   THE KRUSKAL-WALLIS TEST*

The **Kruskal-Wallis test** is a nonparametric alternative to the one-way ANOVA procedure discussed earlier. This test is used when the distributions (for each population) of the variable under consideration have the same shape, but does not require that they be normal or have any other specific shape. Like the previous nonparametric tests we have studied, the Kruskal-Wallis test is based on ranks.

Procedure 16.3 of Section 16.5 outlines the steps for performing a Kruskal-Wallis Test. The assumptions are:

1. Simple random samples

2. Independent samples

3. Same-shape populations

4. All sample sizes are 5 or greater

The TI-83/84 Plus does not have a built-in Kruskal-Wallis test, but the KWTEST program contained in the WeissStats CD will serve the same purpose. This program uses the chi-square distribution to compute the p-value. Therefore, all sample sizes must be 5 or greater. If this is not the case, then the test statistic computed by the program is valid (assuming the number of ties is small), but not the p-value.

Example 16.9   The U.S. Federal Highway Administration conducts annual surveys on motor vehicle travel by type of vehicle and publishes its findings in *Highway Statistics*. Independent random samples of cars, buses, and trucks provided the data on number of miles driven last year, in thousands, shown in Table 16.6. At the 5% significance level, do the data provide sufficient evidence to conclude that a difference exists in last year's mean number of miles driven among cars, buses, and trucks?

Table 16.6	Cars	Buses	Trucks
	19.9	1.8	24.6
	15.3	7.2	37.0
	2.2	7.2	21.2
	6.8	6.5	23.6
	34.2	13.3	23.0
	8.3	25.4	15.3
	12.0		57.1
	7.0		14.5
	9.5		26.0
	1.1		

Solution: Preliminary data analysis (not shown) suggest that the distributions of miles driven have roughly the same shape for cars, buses, and trucks but that those distributions are far from normal

Let μ_1, μ_2, μ_3 denote last year's mean number of miles driven for cars, buses, and trucks, respectively.

Step 1: State the null and alternative hypotheses.
$H_o : \mu_1 = \mu_2 = \mu_3$ (mean miles driven are equal)
H_a : Not all the means are equal.

Step 2: Decide on the significance level, α.
The test is to be performed at the 5% significance level. Thus $\alpha = 0.05$.

Step 3: Compute the value of the test statistic.

The KWTEST program can be used to perform the Kruskal-Wallis test. The program requires that the sample data be entered into List 1, with the number corresponding to the sample located in List 2. For this example, let cars be sample 1, buses sample 2, and trucks sample 3. You may find it easier to enter all the car data into List 1, and then enter the corresponding 1's into List 2 before entering the buses data into List 1. A portion of the list editor with the data entered is shown below in Figure 16.8.

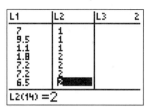
Figure 16.8

1. From the home screen, press PRGM, arrow down to the program KWTEST, and press ENTER. prgmKWTEST will appear on your home screen.

2. Press ENTER to run the program. The results will be displayed. See Figure 16.9.

Figure 16.9

The output displayed is the number of ties, the test statistic H, and the *p*-value. For this example, the number of ties is 2 and the test statistic H = 9.923 when rounded to three decimal places.

Step 4: Obtain the *p*-value.
The *p*-value is 0.007 when rounded to three decimal places. Note that for this example the number of ties was small and all the sample sizes were 5 or greater.

Step 5: If *p*-value $\leq \alpha$, reject H_0; otherwise, do not reject H_0.
Since the *p*-value of 0.007 is less than the specified significance level of 0.05, we reject H_0.

Step 6: Interpret the results of the hypothesis test.
At the 5% significance level, the data provide sufficient evidence to conclude that a difference exists in last year's mean number of miles driven among cars, buses, and trucks.

16.5 Practice Problems

Problem 16.129   Indications are that Americans have become more aware of the dangers of excessive fat intake in their diets, although some reversal of this awareness appears to have developed in recent years. The U.S. Department of Agriculture publishes data on annual consumption of selected beverages in *Food Consumption, Prices and Expenditures*. Independent random samples of lowfat-milk consumptions for 1980, 1995, and 2005 revealed the data in Table 16.7, in gallons. At the 1% significance level, do the data provide sufficient evidence to conclude that there is a difference in mean (per capita) consumption of lowfat milk for the years 1980, 1995, and 2005? Use the Kruskal-Wallis procedure to perform the required hypothesis test.

Table 16.7

1980	1995	2005
11.1	15.5	11.2
10.7	16.0	12.7
8.6	16.1	17.4
9.4	14.7	17.1
9.2	11.5	13.4
15.1	17.1	11.4
11.6	16.2	13.9
8.3		14.6
		15.2

Problem 16.131   The U.S. Bureau of the Census publishes information on the sizes of housing units in *Current Housing Reports*. Independent random samples of single-family detached homes (including mobile homes) in the four U.S. regions yielded the data on square footage in Table 16.8. At the 5% significance level do the data provide sufficient evidence to conclude that a difference exists in median square footage of single-family detached homes among the four U.S. regions? Use the Kruskal-Wallis procedure to perform the required hypothesis test.

Table 16.8

Northeast	Midwest	South	West
3182	2115	1591	1345
2130	2413	1354	694
1781	1639	722	2789
2989	1691	2135	1649
1581	1655	1982	2203
2149	1605	1639	2068
2286	3361	642	1565
1293	2058	1513	1655

Problem 16.150   Use the data (on the WeissStats CD) for the number of three syllable or more words in advertisements from magazines of three different educational levels from Exercise 16.43 of *Introductory Statistics*. At the 5% significance level, do the data provide sufficient evidence of a difference in the mean number of three syllable or more words among the three magazine levels? Use the Kruskal-Wallis procedure to perform the required hypothesis test.

Review
Problem #24   Refer to the data in Review Problem #19. At the 5% significance level, do the data provide sufficient evidence to conclude that a difference in mean losses exists among the three types of robberies? Use the Kruskal-Wallis procedure to perform the required hypothesis test.

Appendix A
Answers to Problems

This appendix contains the answers to the problems in this manual. They are grouped according to the Lessons. On problems where the answer depends on a random sample, all answers will vary. Therefore no specific answer will be given

Lesson 2

2.61

Speed	Rel. Freq.
52-<54	0.057
54-<56	0.143
56-<58	0.171
58-<60	0.229
60-<62	0.2
62-<64	0.086
64-<66	0.057
66-<68	0.029
68-<70	0
70-<72	0
72-<74	0
74-<76	0.029

2.64

Exam Score	Rel. Freq.
30-<40	0.1
40-<50	0
50-<60	0
60-<70	0.15
70-<80	0.15
80-<90	0.4
90-100	0.2

2.66

Heart Rate	Rel. Freq.
50-<60	0.226
60-<70	0.387
70-<80	0.290
80-<90	0.097

2.61 (continued)

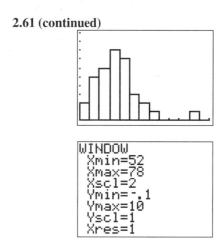

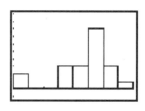

2.64 (continued)

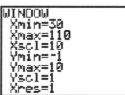

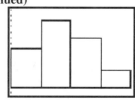

2.66 (continued)

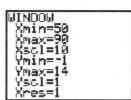

2.52

Review Problem #21

2.53

2.53 (continued)

2.76

2.65

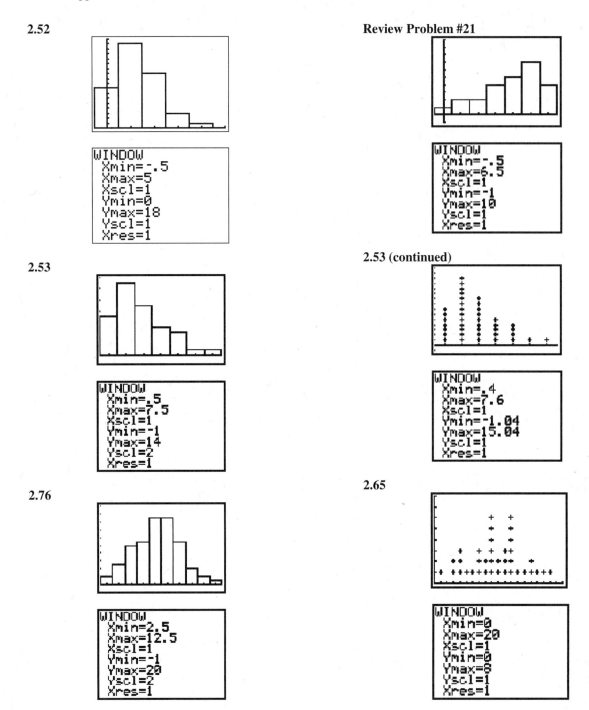

Review Problem #21 (continued)

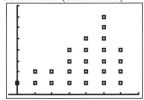

Lesson 3

3.15 mean = 7.3 days median = 6.0 days

3.16 mean = 120.0 kg median = 131.5 kg

3.17 mean = 78.4 tornadoes
          median = 77.0 tornadoes

3.18 mean = 5.70 points
          median = 5.70 points

3.71 range = 6 days
          standard deviation = 2.6 days

3.72 range = 118 kg per hectare per year
          standard deviation = 118 kg per hectare per
              year

3.73 range = 202 tornadoes
          Standard deviation = 53.9 tornadoes

3.74 range = 0.4 points
          Standard deviation = 0.11 points

3.123 Units are in days
          $minX = 1$
          $Q_1 = 4$
          Median = 7
          $Q_3 = 12$
          $maxX = 55$

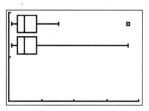

3.124 Units are in thousands of miles.
          $minX = 3.3$
          $Q_1 = 11$
          median = 12.2
          $Q_3 = 14.2$
          $maxX = 16.7$

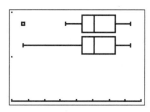

3.125 Units are in kg per hectare per year
          $minX = 57$
          $Q_1 = 88$
          median = 131.5
          $Q_3 = 154$
          $maxX = 175$

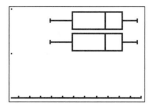

Lesson 4

4.16 a. 0.106   b. 0.748   c. 0.181
          d. 0.488

4.97

	Rookie	1-5	6-10	10+	Total
< 200	0.046	0.062	0.015	0.000	0.123
200-300	0.123	0.185	0.262	0.092	0.662
>300	0.000	0.123	0.092	0.000	0.215
Total	0.169	0.369	0.369	0.092	1.000

4.98

	U.S.	Canada	Mexico	Total
Automobiles	0.534	0.054	0.035	0.623
Motorcycles	0.016	0.001	0.001	0.018
Trucks	0.312	0.029	0.018	0.359
Total	0.862	0.084	0.054	1

4.112

	Fatal	Non-fatal	Total
Australia	0.017	0.104	0.120
Brazil	0.022	0.039	0.061
South Africa	0.015	0.106	0.120
United States	0.009	0.452	0.461
Other	0.067	0.170	0.237
Total	0.130	0.870	1

4.92 (continued) a. 0.84   b. 0.54   c. 0.087   d. 0.021

4.111 a. 0.169   b. 0.123   c. 0.375   d. 0.273

4.112 a. 0.061   b. 0.022   c. 0.364

4.182 657,720

4.185 4896

4.187 5040

Lesson 5

5.7

X	2	3	4	5	6	7	8
P(x)	0.042	0.010	0.021	0.375	0.188	0.344	0.021

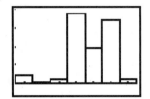 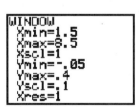

```
WINDOW
Xmin=1.5
Xmax=8.5
Xscl=1
Ymin=-.05
Ymax=.4
Yscl=.1
Xres=1
```

5.8

X	1	2	3	4	5	6	7
P(x)	0.265	0.327	0.161	0.147	0.065	0.022	0.013

5.9   a. 0.991   b. 0.371   c. 0.914   d. 0.557

5.21 a. $\mu = 5.8$ crew members,
b. $\sigma = 1.27$ crew members

5.22 a. $\mu = 2.5$ persons
b. $\sigma = 1.4$ persons

5.23 a. $\mu = 1.9$ color TVs
b. $\sigma = 1.0$ color TVs

5.48 a. 0.410 b. 0.998 c. 0.590

5.63 a. 0.279; 0.685; 0.594
   b. 0.720

5.67
a.

X	P(X=x)
0	0.4861
1	0.3842
2	0.1139
3	0.0150
4	0.0007

b. 0.66 On average, we would expect about 0.66 people of four people under the age of 65 to have no health insurance.

5.84 a. 0.261   b. 0.430

5.85 a. 0.195 b. 0.102 c. 0.704

5.87 a. 0.497   b. 0.966
   c. 0.498

Lesson 6

6.55   a. 0.9875   b. 0.0594
      c. 0.5   d. 0.000032

6.57   a. 0.8577   b. 0.2743
      c. 0.5   d. 0.000013

6.59   a. 0.9105   b. 0.0440
      c. 0.2121   d. 0.1357

6.67 -1.96

6.69 0.67

6.71 -1.645

6.73 0.44

6.75   a. 1.88   b. 2.575

6.77 ± 1.645

6.93   a. 14.66%   b. 31.21%
      c. 16.96 mm; 18.14 mm; 19.32 mm

6.95   a. 73.01%   b. 94.06%
      c. 58.8 minutes   d. 68.6 minutes

6.97   a. 76.47%   b. 0.03%

6.105   a. 0.0085   b. 0.5569

6.22   a. 0.0668 b. 0.1584

6.23   a. 0.7651 b. 0.00

6.24   a. 0.0599 b. 0.2695

6.123

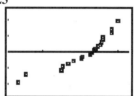

Final exam scores in this introductory statistics course do not appear to be normally distributed.

6.125

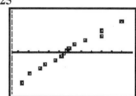

It appears plausible that finishing times for the winners of 1-mile thoroughbred horse races are approximately normally distributed.

6.127

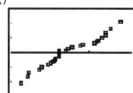

It appears plausible that the average times spent per user per month from January to June of the year in question are approximately normally distributed.

6.145 a. 0.0833   b. 0.4731
c. 0.9370

6.149 a. 0.0478   b. 0.4761
c. 0.0869

Lesson 7
Answers will vary.

Lesson 8

8.31 (63.82, 72.8) We can be 90% confident that the mean number of days to maturity of all short term investments is somewhere between 63.8 and 72.8.

8.32 (860.5, 1034.20) We can be 95% confident that the mean calcium intake of all adults with incomes below the poverty is between 860.5 mg and 1034.2 mg.

8.36 (2.0349, 2.5051) We can be 99% confident that the mean gross earnings of all Rolling Stones concerts in somewhere between $2,034,900 and $2,505,100.

8.42 (44.158, 46.435) We can be 95% confident that the mean weight of all male Ethiopian-born school children, ages
12-15 years old, falls between 44.15 and 46.44 kg.

8.81 a. 1.440   b. 2.447   c. 3.143

8.82 a. 1.740   b. 2.110   c. 2.898

8.96 (182.28, 204.36) We can be 95% confident that the mean cost for a family of four to spend the day at an American amusement park is somewhere between $182.28 and $204.36.

8.103 (14.464, 15.638) We can be 90% confident that the mean depth of all subterranean coruro burrows is somewhere between 14.46 and 15.64 centimeters.

8.106 (22.991, 28.646) We can be 90% confident that the mean commuting time of all local bicycle commuters is somewhere between 22.9 and 28.7 minutes.

Review Problem #16 (81.655, 90.707) We can be 90% confident that the mean arterial blood pressure for all children of diabetic mothers is somewhere between 81.65 and 90.71 millimeters of mercury.

Lesson 9
9.73 H_0: $\mu = 0.5$ H_a: $\mu > 0.5$, $z = 0.24$, $p = 0.4044$. Do not reject H_0. At the 5% significance level, the data fail to provide sufficient evidence to conclude that the mean cadmium level in *Boletus pinicola* mushrooms is greater than the government-recommended limit of 0.5 parts per million.

9.74 H_0: $\mu = \$57.61$ H_a: $\mu \neq \$57.61$, $z = -1.65$, $p = 0.098$. Reject H_0. At the 10% significance level the data provide sufficient evidence to conclude that this year's mean retail price of all history books has changed since the 2005 mean of $57.61.

9.81 H_0: $\mu = 98.6$ H_a: $\mu \neq 98.6$, $z = -7.35$, $p = 0$, At the 1% significance level, the data provides sufficient evidence to conclude that the mean body temperature of healthy humans differs from 98.6°F.

9.101 H_0: $\mu = 4.55$ hr H_a: $\mu \neq 4.55$ hr, $t = .41$, $p = 0.687$. Do not reject H_0. At the 10% significance level, the data do not provide sufficient evidence to conclude that the mean amount of TV watched per day last year differed from 2005.

9.103 H_0: $\mu = 2.3\%$ H_a: $\mu > 2.3\%$, $t = 4.251$, $p = 0.0011$, At the 1% significance level, the data provide sufficient evidence to conclude that the mean available limestone in soil treated with 100% MMBL effluent exceeds 2.3%.

9.104 H_0: $\mu = 1874$ H_a: $\mu \neq 1874$, $t = 2.66$, $p = 0.014$. Reject H_0. At the 5% significance level, the data fail to provide sufficient evidence to conclude that the 2006 mean annual expenditure on apparel and services for consumer units in the Northeast differed from the national mean of $1874.

9.105 H_0: $\mu = 0.9$ H_a: $\mu < 0.9$, $t = -23.703$, $p = 0.000$. Reject H_0. At the 5% significance level, the data provides sufficient evidence to conclude that women with peripheral arterial disease have an unhealthy ABI..

9.107 Yes, it appears reasonable to apply the t-test. The sample size is moderate and the normal probability play show no outliers and is (very roughly linear.

Review Problem #38
H_0: $\mu = 64.5$ lbs H_a: $\mu < 64.5$ lbs,
$t = -1.862$, $p = 0.0351$. Reject H_0. At the 5% significance level, the data provide sufficient evidence to conclude that last year's mean beef consumption is less than the 2002 mean of 64.5 lbs.

9.137 H_0: $\eta = 36.6$ H_a: $\eta > 36.6$,
$W = 33$, $n = 10$ Critical Value = 50. At the 1% significance level, the data fails to provide sufficient evidence to conclude that the median age of today's U.S. residents has increased over the 2007 median age of 36.6 years.

9.139 H_0: $\mu = \$13,015$ H_a: $\mu < \$13,015$,
$W = 13$, $n = 10$, Critical Value= 14. Reject H_0. At the 10 % significance level, the data provide sufficient evidence to conclude that the mean asking price for 2006 Ford Mustang coupes in Phoenix is less than the 2003\9 *Kelley Blue Book* value.

9.140 H_0: $\eta = 7.4$ lb H_a: $\eta \neq 7.4$ lb,
$W = 57.5$, critical values are 17 and 74. Do not reject H_0. At the 5% significance level, the data do not provide sufficient evidence to conclude that this year's median birth weight differs from that in 2002.

9.141 H_0: $\mu = 2.30\%$ H_a: $\mu > 2.30\%$,
$W = 53$, $n = 10$, critical value = 50. Reject H_0. At the 1% significance level, the data provides sufficient evidence to conclude that the mean available limestone in soil treated with 100% MMBL effluent exceeds 2.30%, the percentage ordinarily found.

9.143 H_0: $\mu = 310$ H_a: $\mu < 310$,
$W = 36.5$, $n = 16$, critical value = 36. Do not reject H_0. At the 5% significance level, the data fails to provide sufficient evidence to conclude that the mean content is less than advertised.

Lesson 10

10.39 H_0: $\mu_1 = \mu_2$ H_a: $\mu_1 < \mu_2$, $t = -4.058$, $p = 0.0004$. Reject H_0. At the 5% significance level, the data provides sufficient evidence to conclude that the mean time served for fraud is less than that for firearms offenses.

10.41 H_0: $\mu_1 = \mu_2$ H_a: $\mu_1 > \mu_2$,
$t = 0.520$, $p = 0.3038$. Do not reject H_0. At the 5% significance level, the data fails to provide sufficient evidence to conclude that drinking fortified orange juice reduces PTH level more than drinking unfortified orange juice.

10.51 H_0: $\mu_1 = \mu_2$ H_a: $\mu_1 \neq \mu_2$,
$t = -2.935$, $p = 0.0041$, At the 1% significance level, the data provides sufficient evidence to conclude that the mean daily protein intake of female vegetarians and female omnivores differ.

Review Problem #7 H_0: $\mu_1 = \mu_2$ H_a: $\mu_1 > \mu_2$, $t = 1.539$, $p = 0.0682$. Do not reject H_0. At the 5% significance level, the data fails to provide sufficient evidence to conclude that mean right-leg strength of males exceeds that of females.

10.45 (-12.36, -4.961) We can be 90% confident that the difference between the mean time served for fraud and firearms offenses is somewhere between -12.36 and -4.96.

10.47 (-16.92, 31.72) We can be 90% confident that the difference between the reduction in PTH level of those drinking fortified and unfortified orange juice is somewhere between -16.92 and 31.72 pg/mL.

10.51 (continued) (-18.23, -3.528) We can be 95% confident that the difference between the mean daily protein intakes of female vegetarians and female omnivores is somewhere between -18.23 and -3.53.

Review Problem #8 (-31.2, 599.5) We can be 90% confident that the difference between the mean right leg strength of males and females is somewhere between -31.2 and 599.5.

10.69 H_0: $\mu_1 = \mu_2$ H_a: $\mu_1 \neq \mu_2$, $t = 1.791$,
$p = 0.0794$. Reject H_0. At the 10% significance level, the data provides sufficient evidence to conclude that a difference exists in the mean age at arrest of East German prisoners with chronic PTSD and remitted PTSD.

10.71 H_0: $\mu_1 = \mu_2$ H_a: $\mu_1 < \mu_2$,
$t = -1.651$, $p = 0.0781$. Do not reject H_0. At the 5% significance level, the data fails to provide sufficient evidence to conclude that the mean number of acute postoperative days in the hospital are fewer with the dynamic system than with the static system.

10.73 $H_0: \mu_1 = \mu_2$ $H_a: \mu_1 > \mu_2$,
$t = 3.863$, $p = 0.0006$. Reject H_0. At the 1% significance level, the data provides sufficient evidence to conclude that dopamine activity is higher, on the average, in psychotic patients.

Review Problem #9
$H_0: \mu_1 = \mu_2$ $H_a: \mu_1 < \mu_2$, $t = -4.118$, $p = 0.0001$.
Reject H_0. At the 1% significance level, the data provides sufficient evidence to conclude that, on the average, the number of young per litter of cottonmouths in Florida is less than that of Virginia.

10.75 (0.2, 7.2) We can be 90% confident that the difference between the mean age at arrest of East German prisoners with chronic PTSD and remitted PTSD is somewhere between 0.2 and 7.2 years.

10.77 (-6,97, 0.69) We can be 90% confident that the difference between the mean number of acute postoperative days in the hospital with the dynamic system and the static system is somewhere between –6.97 and 0.69 days.

10.79 (0.00266, 0.01301) We can be 98% confident that the difference between the dopamine activity in psychotic and not psychotic patients is somewhere between 0.00266 and 0.01301.

Review Problem #10 (-3.4, -0.9) We can be 95% confident that the difference between the mean number of young per litter of cottonmouths in Florid and Virginia is somewhere between –3.4 and –0.9.

10.111 $H_0: \mu_1 = \mu_2$ $H_a: \mu_1 < \mu_2$, $M = 33$,
Critical value = 33. Reject H_0. At the 5% significance level, the data fails to provide sufficient evidence to conclude that, in this teacher's chemistry course, students with fewer than two years of high-school algebra have a lower mean semester average than those with two or more years.

10.114 $H_0: \mu_1 = \mu_2$ $H_a: \mu_1 < \mu_2$, $M = 47$,
$M_l = 30$, At the 5% significance level, the data fails to provide sufficient evidence to conclude that the median number of volumes held by public colleges is less than that held by private colleges.

10.115 $H_0: \mu_1 = \mu_2$ $H_a: \mu_1 < \mu_2$,
$M = 65.5$, critical value = 83. Reject H_0. At the 5% significance level, the data provides sufficient evidence to conclude that the mean time served for fraud is less than that for firearms offenses.

10.41 $H_0: \mu_1 = \mu_2$ $H_a: \mu_1 \neq \mu_2$,
$M = 111$, $M_l = 79$, $M_r = 131$, At the 5% significance level, the data fails to provide sufficient evidence to conclude that the mean costs for existing single-family detached homes differ in New York City and Los Angeles.

10.42 $H_0: \mu_1 = \mu_2$ $H_a: \mu_1 < \mu_2$,
$M = 63.5$, $M_l = 35$, At the 5% significance level, the data fails to provide sufficient evidence to conclude that the mean number of acute postoperative days in the hospital are fewer with the dynamic system than with the static system.

10.141 $H_0: \mu_1 = \mu_2$ $H_a: \mu_1 \neq \mu_2$,
$t = 2.148$, $p = 0.0497$. Reject H_0. At the 5% significance level, the data provides sufficient evidence to conclude that the mean height of cross-fertilized and self-fertilized Zea mays differ.

10.143 $H_0: \mu_1 = \mu_2$ $H_a: \mu_1 < \mu_2$,
$t = -4.185$, $p = 0.0004$. Reject H_0. At the 5% significance level, the data provides sufficient evidence to conclude that family therapy is effective in helping anorexic young women gain weight on average.

10.145 $H_0: \mu_1 = \mu_2$ $H_a: \mu_1 > \mu_2$,
$t = 1.053$, $p = 0.1637$. Do not reject H_0. At the 10% significance level, the data fails to provide sufficient evidence to conclude that mean corneal thickness is greater in normal eyes than in eyes with glaucoma.

10.147 (0.03119, 41.835) We can be 95% confident that the difference between the mean height of cross-fertilized and self-fertilized Zea mays is somewhere between 0.031 and 41.84 eighths of an inch.

10.149 (-10.30, -4.23) We can be 90% confident that the difference between the mean weights of anorexic young women before and after receiving family-therapy treatment is somewhere between –10.30 and –4.23 lbs.

10.151 (-1.4, 9.4) We can be 80% confident that the difference between the mean corneal thickness of normal eyes and eyes with glaucoma is somewhere between –1.4 and 9.4 microns.

10.173 H_0: $\mu_1 = \mu_2$ H_a: $\mu_1 \neq \mu_2$, W = 96, n = 15, Critical values are 16 and 104. Do not reject H_0. At the 1% significance level, the data do not provide sufficient evidence to conclude that the mean heights of cross-fertilized and self-fertilized Zea mays differ.

10.175 H_0: $\mu_1 = \mu_2$ H_a: $\mu_1 < \mu_2$, W = 11, n = 17, critical value = 41. Reject H_0. At the 5% significance level, the data provide sufficient evidence to conclude that family therapy is effective in helping anorexic young women gain weight.

10.177 H_0: $\mu_1 = \mu_2$ H_a: $\mu_1 > \mu_2$, W = 20.5, n = 7, critical value = 22. Do not reject H_0. At the 10% significance level, the data fails to provide sufficient evidence to conclude that mean corneal thickness is greater in normal eyes than in eyes with glaucoma.

Review Problem #14 H_0: $\mu_1 = \mu_2$ H_a: $\mu_1 > \mu_2$, W = 46, n = 10, Critical value = 44. Reject H_0. At the 5% significance level, the data provide sufficient evidence to conclude that, on average, the eyepiece method gives a greater fiber-density reading than the TV-screen method.

Lesson 11

11.5 a. 32.852   b. 10.117

11.7 a. 18.307   b. 3.247

11.9 a. 1.646   b. 15.507

11.11 a. 0.831, 12.833
   b. 13.844, 41.923

11.23 H_0: $\sigma = 0.27$ H_a: $\sigma > 0.27$, $\chi^2 = 70.361$, p ≈ 0, At the 1% significance level, the data provides sufficient evidence to conclude that the process variation for this piece of equipment exceeds the analytical capability of 0.27.

11.25 H_0: $\sigma = 0.2$ fl. oz. H_a: $\sigma < 0.2$ fl. oz., $\chi^2 = 8.317$, p = 0.128, At the 5% significance level, the data fails to provide sufficient evidence to conclude that the standard deviation of the amounts being dispensed is less than 0.2 fl. oz.

Review Problem #9 H_0: $\sigma = 16$ H_a: $\sigma \neq 16$, $\chi^2 = 21.110$, p = 0.736, At the 10% significance level, the data does not provide sufficient evidence to conclude that IQs measured on this scale have a standard deviation different from 16 points.

11.29 (0.487, 1.570) We can be 98% confident that the process variation of the piece of equipment under consideration is somewhere between 0.487 and 1.570.

11.31 (0.119, 0.225) We can be 90% confident that the standard deviation of the amounts being dispensed in the coffee cups is somewhere between 0.119 and 0.225.

Review Problem # 10 (11.717, 20.875) We can be 95% confident that the standard deviation of the IQs measured on the Stanford Revision of the Binet-Simon Intelligence Scale is somewhere between 11.7 and 20.9.

11.53 a. 1.89
   b. 2.47
   c. 2.14

11.55 a. 2.88
   b. 2.10
   c. 1.78

11.57 a. 0.12
   b. 3.58

11.59 a. 0.18, 9.07
   b. 0.33, 2.68

11.69 H_0: $\sigma_1 = \sigma_2$ H_a: $\sigma_1 > \sigma_2$, F = 2.185, p = 0.0348, At the 5% significance level, the data provides sufficient evidence to conclude that there is less variation among final-exam scores when the new teaching method is used.

11.71 H_0: $\sigma_1 = \sigma_2$ H_a: $\sigma_1 \neq \sigma_2$, F = 1.219, p = 0.6241, At the 10% significance level, the data fails to provide sufficient evidence to conclude that variation in anxiety-test scores differs between patients who are shown videotapes of progressive relaxation exercises and those who are shown neutral tapes.

11.75 (1.04, 2.01) We can be 90% confident that the ratio of the population standard deviations of final exam scores for students taught by the conventional method and those taught by the new method is somewhere between 1.04 and 2.01. (i.e. $1.04\,\sigma_2 < \sigma_1 < 2.01\,\sigma_2$). In other words, we can be 90% confident that the standard deviation of final exam scores for students taught by the conventional method is somewhere between 1.04 and 2.01 times greater than that for those taught by the new method.

11.77 (0.79, 1.52) We can be 90% confident that the ratio of the population standard deviations of scores for patients who are shown videotapes of progressive relaxation exercises and those who are shown neutral videotapes is somewhere between 0.79 and 1.52. (i.e. $0.79\,\sigma_2 < \sigma_1 < 1.52\,\sigma_2$). In other words, we can be 90% confident that the standard deviation of scores for patients who are shown videotapes of progressive relaxation exercises is somewhere between 1.27 times less than and 1.52 times greater than that for those who are shown neutral videotapes.

11.81 H_0: $\sigma_1 = \sigma_2$ H_a: $\sigma_1 \neq \sigma_2$, F = 0.927, p = 0.750, At the 5% significance level, the data fail to provide sufficient evidence to conclude that variation in skull measurements differs between the two populations.

Lesson 12

12.13 (0.7845, 0.7955) We can be 95% confident that the percentage of Americans who believe states should be allowed to conduct random drug testing on elected officials is somewhere between 78.45 and 79.55%.

12.26 (0.6257, 0.6738) We can be 95% confident that between 62.5 and 67.4% of all Americans drink beer, wine, or hard liquor at least occasionally.

12.28 (0.240, 0.399) We can be 90% confident that the percentage of Malaysians infected with the Nipah virus who will die from encephalitis is somewhere between 24 and 39.9%.

12.69 H_0: p = 0.72 H_a: p < 0.72, z = -4.93, p ≈ 0, At the 5% significance level, the data provide sufficient evidence to conclude that the percentage of Americans who approve of labor unions now has decreased since 1936.

12.65 H_0: p = 0.5 H_a: p > 0.5, z = 2.33, p = 0.0098, At the 5% significance level, the data provides sufficient evidence to conclude that a majority of Generation Y users use the Internet to download music.

12.67 H_0: p = 0.128 H_a: p ≠ 0.128, z = 3.41, p ≈ 0, At the 10% significance level, the data provides sufficient evidence to conclude that the percentage of 18-25 year olds who currently use marijuana or hashish has changed from the 2000 percentage of 13.6%.

12.68 H_0: p = 0.098 H_a: p < 0.098, z = -2.39, p = 0.008, At the 1% significance level, the data provide sufficient evidence to conclude that in 2006, families with incomes below the poverty was lower among those living in Wyoming than the national percentage of 9.8%.

12.89 H_0: $p_1 = p_2$ H_a: $p_1 < p_2$ z = -2.61, p = 0.0045, At the 1% significance level, the data provides sufficient evidence to conclude that women who take folic acid are at lesser risk of having children with major birth defects.

12.91 H_0: $p_1 = p_2$ H_a: $p_1 \neq p_2$ z = -1.52, p = 0.1285, At the 10% significance level, the data fails to provide sufficient evidence to conclude that there is a difference in seat-belt usage between drivers 25-34 years old and those 45-64 years old.

12.93 H_0: $p_1 = p_2$ H_a: $p_1 > p_2$ z = 1.41, p = 0.0795, At the 5% significance level, the data fails to provide sufficient evidence to conclude that the percentage who are overweight is greater for those whose highest degree is a Bachelors than for those with a graduate degree.

Review Problem #14

H_0: $p_1 = p_2$ H_a: $p_1 < p_2$ z = -4.17, p ≈ 0, At the 1% significance level, the data provides sufficient evidence to conclude that the percentage of Maricopa County residents who thought the state's economy would improve over the next two years was less during the time of the first poll than during the time of the second poll.

12.95 (-0.0191, -0.0007) We can be 98% confident that the difference between the rates of major birth defects for babies born to women who have taken folic acid and those born to women who have not is somewhere between -0.0191 and -0.0007.

12.97 (-0.0624, 0.0024) We can be 90% confident that the difference between the proportions of seat belt users for drivers in the age groups 25-34 years and 45-64 years is somewhere between -0.0624 and 0.0024.

12.99 (-0.0159, 0.0972) We can be 95% confident that the difference between the proportions of overweight adults whose highest degree is a Bachelors and those with a graduate degree is somewhere between –0.0159 and 0.0972.

Chapter 12 Review #15 (-0.1936, -0.0464) We can be 99% confident that the difference between the proportion of Maricopa County residents who thought the state's economy would improve over the next two years in the first poll and the second poll is somewhere between -0.1936 and -0.0464.

Lesson 13

13.6 a. 32.852   b. 10.117

13.7 a. 18.307   b. 3.247

13.27 H_0: This year's distribution of political views for incoming college freshmen is the same as the 2000 distribution. H_a: This year's distribution of political views for incoming college freshmen differs from the 2000 distribution. $\chi^2 = 4.667$, p = 0.0969, At the 5 % significance level, the data fails to provide sufficient evidence to conclude that this year's distribution of political views for incoming college freshmen differs from the 2000 distribution.

13.29 H_0: The color distribution of M&Ms is the same as that reported by M&M/Mars consumer affairs. H_a: The color distribution of M&Ms is different from that reported by M&M/Mars consumer affairs. $\chi^2 = 4.091$, p = 0.5364, At the 5% significance level, the data fails to provide sufficient evidence to conclude the color distribution of M&Ms is different from that reported by M&M/Mars consumer affairs.

13.31 H_0: The die is not loaded. H_a: The die is loaded. $\chi^2 = 2.48$, p = 0.7795, At the 5% significance level, the data fails to provide sufficient evidence to conclude that the die was loaded.

13.71 H_0: There is no association between the ratings of Siskel and Ebert. H_a: There is an association between the ratings of Siskel and Ebert. $\chi^2 = 45.357$, p ≈ 0, At the 1% significance level, the data provide sufficient evidence to conclude that there is an association between the ratings of Siskel and Ebert.

13.73 H_0: Social class and the frequency of playing "I Spy" games are not associated. H_a: Social class and the frequency of playing "I Spy" games are associated. $\chi^2 = 8.715$, p = 0.0128, At the 5% significance level, the data provides sufficient evidence that Social class and the frequency of playing "I Spy" games are associated.

13.75 H_0: Size of city and status in practice are statistically independent for U.S. lawyers. H_a: Size of city and status in practice are statistically dependent for U.S. lawyers. $\chi^2 = 6.475$, p = 0.372, At the 5% significance level, the data fails to provide sufficient evidence to conclude that size of city and status in practice are statistically dependent for U.S. lawyers. Note: 3/12 or 25% of the expected frequencies are less than 5 which violates an assumption for this test.

13.92 H_0: Residents of the red, blue, and purple states are homogeneous with respect to their view. H_a: Residents of the red, blue, and purple states are nonhomogeneous with respect to their view. $\alpha = 0.05$; $\chi 2 = 15.346$; critical value = 12.592; *P-value* = 0.018; reject H_0; at the 5% significance level, the data provide sufficient evidence to conclude that the residents of the red, blue, and purple states are nonhomogeneous with respect to their view.

13.96 H_0: $p_1 = p_2$; Ha: $p_1 = p_2$; $\alpha = 0.05$; $\chi 2 = 0.827$; *p-value* = 0.363; do not reject H_0; at the 5% significance level, the data do not provide sufficient evidence to conclude that a difference exists in the approval percentages of all U.S. adults between the two months.

Lesson 14

14.51

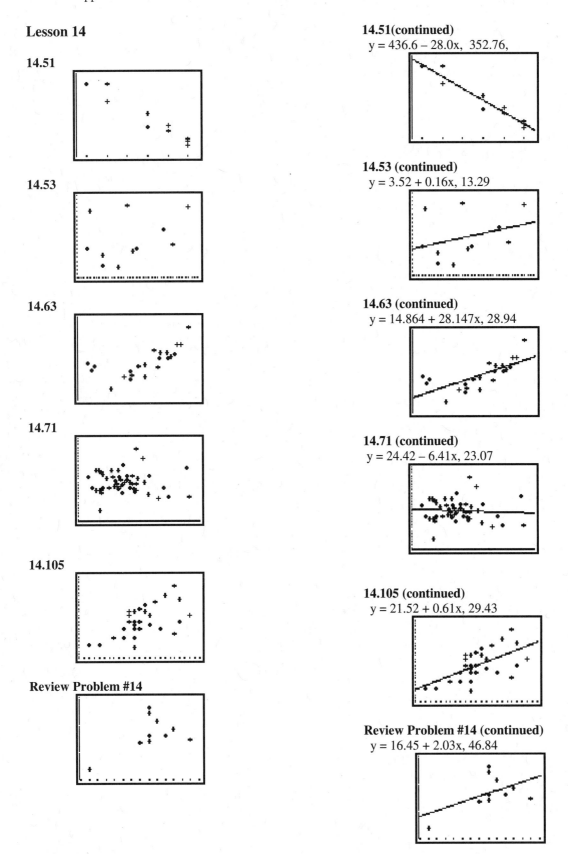

14.51(continued)
y = 436.6 – 28.0x, 352.76,

14.53

14.53 (continued)
y = 3.52 + 0.16x, 13.29

14.63

14.63 (continued)
y = 14.864 + 28.147x, 28.94

14.71

14.71 (continued)
y = 24.42 – 6.41x, 23.07

14.105

14.105 (continued)
y = 21.52 + 0.61x, 29.43

Review Problem #14

Review Problem #14 (continued)
y = 16.45 + 2.03x, 46.84

14.89 25794, 24158.45, 1635.55

14.91 296.6819, 32.517, 264.165

14.101 4.1616, 3.3709, 0.7907

14.105 674, 250.57, 423.43

Review Problem #15
1384.5, 361.66, 1022.84

14.89 (continued) $r^2 = 0.937$

14.91 (continued) $r^2 = 0.110$

14.101 (continued) $r^2 = 0.005$

14.105 (continued) $r^2 = 0.372$

Review Problem #15 (continued)
$r^2 = 0.261$

14.123 $r = -0.968$

14.125 $r = 0.331$

14.139 $r = -0.069$

14.143 $r = 0.610$

Review Problem #16 $r = 0.511$

Lesson 15

15.23 14.30   **15.25** 5.42

15.37 3.02   **15.38** 3.82

Review Problem #12 11.31

15.23 (continued)

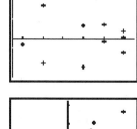

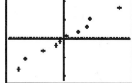

15.25(continued)

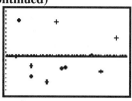

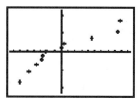

15.37(continued)

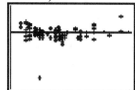

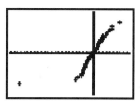

15.38(continued)

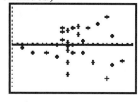

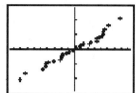

Review Problem # 13

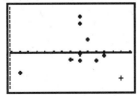

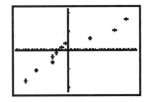

15.51 H_0: $\beta_1 = 0$ H_a: $\beta_1 \neq 0$,
t = -10.870, p ≈ 0, At the 5% significance level, the data provides sufficient evidence to conclude that age is useful as a predictor of price for Corvettes.

15.53 H_0: $\beta_1 = 0$ H_a: $\beta_1 \neq 0$, t =1.053, p = 0.3200, At the 5% significance level, the data fails to provide sufficient evidence to conclude that weight is useful as a predictor of quantity of volatile compounds emitted for potato plants.

15.72 H_0: $\beta_1 = 0$ H_a: $\beta_1 \neq 0$, t = 5.191, p = 0.0001, At the 5% significance level, the data provides sufficient evidence to conclude that gas mileage is useful as a predictor of engine displacement.

15.69 H_0: $\beta_1 = 0$ H_a: $\beta_1 \neq 0$,
t = -12.353, p ≈ 0, At the 5% significance level, the data provides sufficient evidence to conclude that engine displacement is a useful predictor of engine displacement.

15.70 H_0: $\beta_1 = 0$ H_a: $\beta_1 \neq 0$, t = 4.143, p= 0.0003, At the 5% significance level, the data provides sufficient evidence to conclude that estriol levels of pregnant women are useful as a predictor of birth weight of children.

Review Problem #14 H_0: $\beta_1 = 0$ H_a: $\beta_1 \neq 0$, t = 1.682, p = 0.1311, At the 5% significance level, the data fails to provide sufficient evidence that student-to-faculty ratio is useful as a predictor of graduation rate.

15.57 (-36.59, -19.33)

15.59 (-0.1901, 0.5102)

Review Problem #14 (continued)
(-0.2112, 4.2713)

15.109 H_0: $\rho = 0$ H_a: $\rho < 0$, t = -10.870, p-value ≈ 0, At the 10% significance level, the data provides sufficient evidence to conclude that age and price are negatively linearly correlated.

15.111 H_0: $\rho = 0$ H_a: $\rho > 0$, t = 1.053, p-value = 0.1600, At the 5% significance level, the data fails to provide sufficient evidence to conclude that plant weight and quantity of volatile compounds emitted for potato plants are positively linearly correlated.

15.123 H_0: $\rho = 0$ H_a: $\rho \neq 0$,
t = -12.353, p-value ≈ 0, At the 1% significance level, the data provides sufficient evidence to conclude that gas mileage and engine displacement are linearly correlated.

15.124 H_0: $\rho = 0$ H_a: $\rho \neq 0$, t = 4.143, p-value = 0.0003, At the 10 % significance level, the data provides sufficient evidence to conclude that estriol levels of pregnant women and birth weights of children are linearly correlated.

Review Problem #16
H_0: $\rho = 0$ H_a: $\rho > 0$, t = 1.682,
p-value = 0.0655, At the 5% significance level, the data fails to provide sufficient evidence to conclude that student-to-faculty ratio and graduation rate data are positively linearly correlated.

15.131 H_0: Final exam scores are normally distributed. H_a: Final exam scores are not normally distributed. Rp = 0.939 < 0.951, At the 5% significance level, the data provides sufficient evidence to conclude final exam scores are not normally distributed.

15.133 H_0: The finishing times are normally distributed. H_a: The finishing times are not normally distributed. Rp = 0.989 > 0.905. At the 1% significance level, the data fails to provide sufficient evidence to conclude that the finishing times for winners of 1-mile thoroughbred horse races are not normally distributed.

Lesson 16

16.7 a. 1.89   b. 2.47   c. 2.14

16.8 a. 9.89   b. 4.68   c. 13.38

16.9 a. 2.88   b. 2.10   c. 1.78

16.10 a. 3.22   b. 5.39   c. 4.07

16.49 H_0: $\mu_1 = \mu_2 = \mu_3$ (mean number of copepods are equal) H_a: Not all means are equal. F = 54.576, p ≈ 0, At the 5% significance level the data provides sufficient evidence to conclude that a difference exists in mean number of copepods among the three different diets.

16.70 H_0: $\mu_1 = \mu_2 = \mu_3$ (mean number of three syllable or more words are equal) H_a: Not all means are equal. F = 1.921, p = 0.1569, At the 5% significance level the data fails to provide sufficient evidence to conclude that there is a difference in the mean number of three syllable or more words among the three magazine levels.

Review Problem #21 H_0: $\mu_1 = \mu_2 = \mu_3$ (mean losses are the same) H_a: Not all means are equal. F = 16.60, p–value =0.000, At the 5% significance level, the data provides sufficient evidence to conclude that a difference in mean losses exists among the three types of robberies.

16.95

	D	B
B	(102.2, 199.4)	
M	(114.6, 211.9)	(-36.0, 61.1)

293.75   306.25   457

We can be 95% confident that the diatom diet resulted in a higher mean than either the bacteria or the macroalgae diets.

16.116

	1	2
2	(-1.6, 15.6)	
3	(-4.7, 12.5)	(-11.6, 5.5)

11.2   14.3   18.1

We can be 95% confident that there is no difference in the mean number of three syllable or more words among the three magazine levels.

Review Problem # 23

	1	2
2	(-383.3, 5.2)	
3	(24.4, 412.9)	(222.4,592.9)

582.17   695.8   824.8

We can be 95% confident that the mean losses of convenience store robberies are less than at highway and gas stations.

16.129 H_0: $\mu_1 = \mu_2 = \mu_3$ (mean consumptions are equal) H_a: Not all means are equal. H = 0.19, p = 0.909, Do not reject H_0 At the 1% significance level, the data do not provide sufficient evidence to conclude that there is a difference in mean (per capita) consumption of lowfat milk for the years 1980, 1995, and 2005.

16.131 H_0: $\mu_1 = \mu_2 = \mu_3 = \mu_4$ (median square footage are equal) H_a: Not all means are equal. H = 6.369, p = 0.0950, At the 5% significance level, the data fails to provide sufficient evidence to conclude that a difference exists in median square footage of single-family detached homes among the four U.S. regions.

16.150 H_0: $\mu_1 = \mu_2 = \mu_3$ (mean number of three syllable or more words are equal) H_a: Not all means are equal. H = 4.349, p = 0.1137, At the 5% significance level, the data fails to provide sufficient evidence to conclude that there is a difference in the mean number of three syllable or more words among the three magazine levels.

Review Problem #24 H_0: $\mu_1 = \mu_2 = \mu_3$ (mean losses are the same) H_a: Not all means are equal. H = 6.819, p = 0.0331, At the 5% significance level, the data provides sufficient evidence to conclude that a difference in mean losses exists among the three types of robberies.